Autorenkollektiv

Analytical Chemistry Progress

Autorenkollektiv

Analytical Chemistry Progress

With 74 Figures and 18 Tables

AKADEMIE-VERLAG · BERLIN
1985

Die Originalausgabe erscheint im Springer-Verlag
Berlin · Heidelberg · New York · Tokyo
in der Schriftenreihe „Topics in Current Chemistry“,
Volume 126

Vertrieb ausschließlich für die DDR und die sozialistischen Länder

Erschienen im Akademie-Verlag, DDR-1086 Berlin, Leipziger Straße 3—4
Lizenznummer: 202 · 100/499/85
Printed in the German Democratic Republic
Gesamtherstellung: VEB Druckerei „Thomas Müntzer“, 5820 Bad Langensalza
Umschlaggestaltung: Karl Salzbrunn
LSV 1235
Bestellnummer: 763 4369 (6874)
06800

Table of Contents

Laser-Enhanced Ionization Spectrometry
R. B. Green 1

Laser Spectroscopy of Biomolecules
A. Anders 23

Ion Pair Chromatography on Reversed-Phase Layers
D. G. Volkmann 51

Evaluation and Calibration in Quantitative Thin-Layer Chromatography
S. Ebel 71

Evaluation Systems in Quantitative Thin-Layer Chromatography
I. M. Böhrer 95

Author Index Volumes 101–126 119

Laser-Enhanced Ionization Spectrometry

Robert B. Green

Department of Chemistry, University of Arkansas, Fayetteville, Arkansas 72701, USA

Table of Contents

1 Introduction 2

2 Signal Production and Collection 7
2.1 Ion Production 7
2.2 Ion Collection 8

3 Analytical Considerations 14
3.1 Instrumentation 14
3.2 Methodology 17
3.3 Limits to Sensitivity 18

4 Applications 18
4.1 General 18
4.2 Flame Diagnostics 19

5 The Future of LEI 20

6 Acknowledgments 21

7 References 21

Laser-enhanced ionization (LEI) is one of a family of laser-induced ionization techniques which have been exploited for analytical spectrometry. In LEI, a pulsed dye laser is used to promote analyte atoms to a bound excited state from which they are collisionally ionized in a flame. The resulting current is detected with electrodes and is a measure of the concentration of the absorbing species. LEI may proceed by photoexcitation (via one or more transitions) and thermal ionization or a combination of thermal excitation, photoexcitation, and thermal ionization. LEI detection limits are competitive with — and in many cases superior to — those obtained with other techniques of atomic spectrometry. This chapter will provide an introduction to LEI spectrometry and its capabilities. Signal production and collection will be discussed along with the more practical aspects of LEI spectrometry. The applications of LEI spectrometry will be reviewed and the future of LEI spectrometry will be assessed.

1 Introduction

Since their invention in 1960, lasers have made a significant impact in chemistry. Fundamental and applied spectroscopy have been major benefactors of the new laser technology. Analytical laser spectrometry is currently a vital and growing field of research.

Laser-enhanced ionization (LEI) is one of a family of laser-induced ionization techniques which have been exploited for analytical spectrometry. The laser-induced ionization schemes which are important for flame spectrometry are illustrated in Fig. 1.

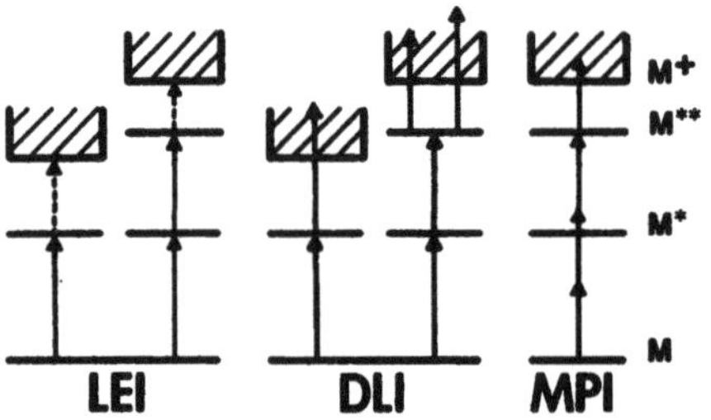

Fig. 1. Laser-induced ionization schemes in flames. LEI-laser-enhanced ionization, DLI-dual laser ionization, MPI-multiphoton ionization

LEI utilizes a pulsed dye laser to promote analyte atoms to a bound excited state. Laser excitation enhances the thermal (collisional) ionization rate of the analyte atom, producing a measurable current in the flame [1,2]. The laser-related current is detected with electrodes and is a measure of the concentration of the absorbing species. LEI may proceed by photoexcitation (via one or more transitions) and thermal ionization or a combination of thermal excitation, photoexcitation, and thermal ionization.

Resonance ionization spectrometry (RIS) is another laser-induced ionization technique which has been applied to the analytical determination of trace metals. Although RIS was developed for low background sample reservoirs, its high sensitivity and relation to the laser-induced flame techniques make it worthy of discussion. RIS involves laser excitation of one or more bound transitions, followed by **direct** laser photoionization of the analyte from the excited state [3-5]. Laser excitation and photoionization may proceed by stepwise or multiphoton processes or a combination of both. Stepwise processes occur as a result of the absorption of two or more photons, usually of different energies, via common intermediate states. Multiphoton excitation refers to the simultaneous absorption of several photons of identical or different energy via virtual intermediate states [6]. Multiphoton excitation may be accomplished with a single laser in some cases. Because of the low probability of multiphoton transitions, stepwise excitation generally yields lower limits of detection. RIS does not rely on thermal energy to promote ionization or excitation, as does LEI. In fact, the single-atom detection limits reported with RIS would not be possible without a low background atom reservoir. Two 455.5 nm photons from a single dye laser have been used to produce a free electron from a single cesium atom by two-photon absorption [7]. The electron was detected in a gas proportional counter. The cesium atom was generated and ejected into the detection volume by the fission of ^{252}Cf.

Low background sampling schemes, the high signal collection efficiency available with existing ion detectors, and the capability of lasers for saturation of atomic transitions have contributed to the excellent sensitivity provided by RIS. Collisionally-assisted RIS and stepwise LEI are essentially synonomous [8].

Because a low background sample reservoir is required to achieve the maximum sensitivity with RIS, most of the analytical applications to date, although important, have been rather specialized [3-5]. Some progress has been made in the adaptation of RIS to real samples. Recently, RIS has been used to determine trace sodium impurities in vaporized single-crystal silicon produced by laser ablation of the solid [9]. This application required three lasers: the ablation laser and two RIS probe lasers, tuned to different wavelengths. A single wavelength two-photon absorption process was dimissed because it required at least two orders-of-magnitude higher laser intensity to produce the same detection limits and a greater possibility of a spurious ionization background existed.

Although the sensitivity obtained with RIS is impressive, analytical calibration in specialized reservoirs presents some practical problems. In the laser ablation/RIS scheme, these limitations are particularly pronounced because of the nature of the sampling process. Relative calibration techniques, commonly used by analytical chemists, imply the availability of accurate standards at the appropriate concentration [9]. The current state of knowledge about the dynamics of the ablation process did not permit absolute calibration with any confidence. The coupling of highly sensitive and selective RIS with mass spectrometry should result in more generally useful applications to analytical chemistry [10].

Purely optical techniques, such as *laser-induced fluorescence* (LIF), have also approached the limit of single atom detection in a laser excitation volume [11,12]. The difference in RIS and LIF is that in the latter, the single atom is recycled through the excited state multiple times to generate a detectable time-averaged photon flux. Thus, the photomultiplier has not achieved the same level of detection efficiency as the proportional counter [13]. Again, single atoms may be detected only under the most favorable conditions. The LIF experiments have used a sodium vapor cell as the source of atoms. The analyte atoms are confined to a small volume and the cells have been constructed to minimize scattered excitation light, a limiting factor for resonance LIF experiments. Fluorescence quenching environments must also be avoided. Single atom detection is an excellent goal for analytical spectroscopists but realistically other factors must be considered.

Dual-laser ionization (DLI) draws from both LEI and RIS [14]. DLI utilizes the flame sample reservoir common to LEI and photoexcitation schemes which are similar to RIS. DLI may be viewed as an extension of LEI or it could be referred to as RIS in flames.

Since the second laser ionizes the excited atom in DLI, this step may be accomplished by an off-resonant photon (see Fig. 1). If a nitrogen laser-pumped dye laser provides the resonant photon, a fraction of the nitrogen laser beam can conveniently ionize the atom from the laser-excited state [14,15]. It is theoretically possible to photoionize every atom in the periodic table except helium and neon using five RIS ionization schemes involving stepwise and/or multiphoton excitation [4]. Presumably these schemes will also be useful for DLI. However it is not reasonable to expect that DLI (or LEI) are capable of producing similar detection limits because of the differences in the sample reservoir.

The DLI detection limits reported to date [14] are several orders-of-magnitude higher than published LEI detection limits [2]. DLI should produce lower detection limits than LEI in some sample reservoirs. If a low temperature or low collision sample cell (e.g., a flame which excludes molecular nitrogen) is used, photoionization should predominate over collisional ionization from the same excited state. DLI may also be preferable to single-wavelength LEI when the energy defect between the laser-excited state and the ionization potential is large. Atomic hydrogen [16] and atomic oxygen [17] have been detected in hydrogen-air and hydrogen-oxygen-argon atmospheric pressure flames, respectively, using resonantly-enhanced multiphoton ionization detected with biased electrodes. Since the first excited states of both hydrogen and oxygen are approximately 10 eV above the ground state, conventional optical detection is difficult. The excited states were populated by two-photon absorption while ionization was accomplished with a third photon. In the hydrogen experiment, a 266 nm photon (4th harmonic of the Nd:YAG laser) and a 224 nm photon (frequency-doubled dye laser summed with the YAG fundamental) were resonant with the two photon transition. One photon from either beam was sufficiently energetic to ionize the excited atom. The two photon transition in oxygen was excited by two 226 nm photons from the frequency-doubled and mixed dye laser output while a third 226 nm photon ionized the atom. Even under conditions where photoionization would be expected to predominate, single-wavelength LEI using a nonresonance (i.e., thermally-populated) transition or stepwise LEI may produce detection limits which are equivalent to DLI.

In the more common situation for spectrochemical analysis, where an acetylene-air flame is the sample reservoir, there may be no sensitivity advantage to using a second (or third) laser to photoionize the excited analyte atom. Stepwise excitation schemes also increase the efficacy of thermal ionization [18,19]. Two cases in which DLI was ultimately abandoned in favor of a stepwise LEI excitation scheme illustrate the predominance of collisional ionization when the energy defect is relatively small. The second harmonic of a Nd:YAG laser was used to pump two dye lasers which excited sequential sodium transitions at 589.0 and 568.8 nm, respectively [20]. In addition, the 1064 nm YAG fundamental was introduced into the excitation volume in the flame to directly photoionize the excited sodium atoms. The third excitation step did not increase the laser-related signal. Subsequent measurements used only the dye lasers for stepwise excitation to a bound excited state with LEI detection. In a similar experiment, the YAG second harmonic was directed into a flame atom reservoir in addition to pumping two lasers tuned to the 670.8 (λ_1) and 610.4 nm (λ_2) lithium transitions, respectively [21]. Again the third step provided by the YAG laser

▶

Fig. 2. LEI periodic chart of the elements, indicating experimental limits of detection (in ng/ml) NOTE: In some instances, ml has been changed to ml. I assume that ml is the form you neter and the excitation scheme for the elements observed to date. R = resonant; N = nonresonant; S = stepwise, resonant; NS = nonresonant, stepwise. Other elements shown are expected to yield LEI signals in flames. Omitted elements are not amenable to flame spectrometry. Detection limits were obtained from the following sources: Na, K (38); Al, Sc, Ti, V, Y, Tm, Lu (39); Rb, Cs (57); all other elements [2]

LEI PERIODIC TABLE

1																1	2
3 *Li* 0.001 R	4 Be											5 B	6	7	8	9	10
11 *Na* 0.03 R	12 *Mg* 0.1 R											13 *Al* 0.1 N	14 Si	15 P	16 S	17	18
19 *K* 0.1 R	20 *Ca* 0.1 N	21 *Sc* 0.2 N	22 *Ti* 600 N	23 *V* 2 N	24 *Cr* 2 N	25 *Mn* 0.02 S	26 *Fe* 2 R	27 *Co* 0.08 S	28 *Ni* 0.08 NS	29 *Cu* 0.07 S	30 Zn	31 *Ga* 0.07 R	32 Ge	33 As	34 Se	35	36
37 *Rb* 0.1 R	38 *Sr* 4 R	39 *Y* 10 N	40 Zr	41 Nb	42 Mo	43 Tc	44 Ru	45 Rh	46 Pd	47 *Ag* 1 R	48 *Cd* 0.1 S	49 *In* 0.006 R	50 *Sn* 0.3 NS	51 Sb	52 Te	53 I	54
55 *Cs* 0.004 R	56 *Ba* 0.2 R	57 La	72 Hf	73 Ta	74 W	75 Re	76 Os	77 Ir	78 Pt	79 *Au* 1 S	80 Hg	81 *Tl* 0.09 N	82 *Pb* 0.09 S	83 *Bi* 2 R	84	85	86
87	88	89	104														

58 Ce	59 Pr	60 Nd	61	62 Sm	63 Eu	64 Gd	65 Tb	66 Dy	67 Ho	68 Er	69 *Tm* 200 R	70 Yb	71 *Lu* 0.2 N
90 Th	91	92 U	93	94	95	96	97	98	99	100	101	102	103

was expected to photoionize the excited lithium. As before, this radiation had no effect on the ionization signal attesting to the efficiency of the collisional ionization processes in the flame which are exploited in LEI spectrometry.

LEI and DLI have similar methodologies, require similar instrumentation and are complementary techniques. Single-wavelength LEI is the least complex and expensive experiment, followed by DLI using a dye laser plus a fraction of the pump laser beam for excitation. Stepwise LEI schemes generally will require two tunable dye lasers pumped by a third laser. Stepwise LEI provides additional selectivity over DLI which can be essential for analysis of real samples. A two dye laser system will also permit optimization of a DLI signal by tuning the second laser to the exact energy required for ionization.

Multiphoton ionization (MPI) of atoms can also occur in flames [22] but it has not been shown to be a viable excitation scheme for trace metal determination. MPI has been used effectively for spectroscopy of organic molecules [23]. MPI typically proceeds via the simultaneous absorption of three or more photons with a single intermediate state (see Fig. 1). In flames, nonresonant MPI can contribute to the background due to the ionization of any species present. A background current may be observed due to the MPI of added and native flame species by high power, pulsed laser-pumped dye lasers. This additive signal may be compensated for by scanning the laser wavelength across the atomic line or eliminated by reducing the laser power or expanding the laser beam.

As this discussion suggests, LEI spectrometry shares many of the properties of other atomic spectroscopic techniques while possessing unique features which complement or supersede other methods. Since a laser is the excitation source, no spectral

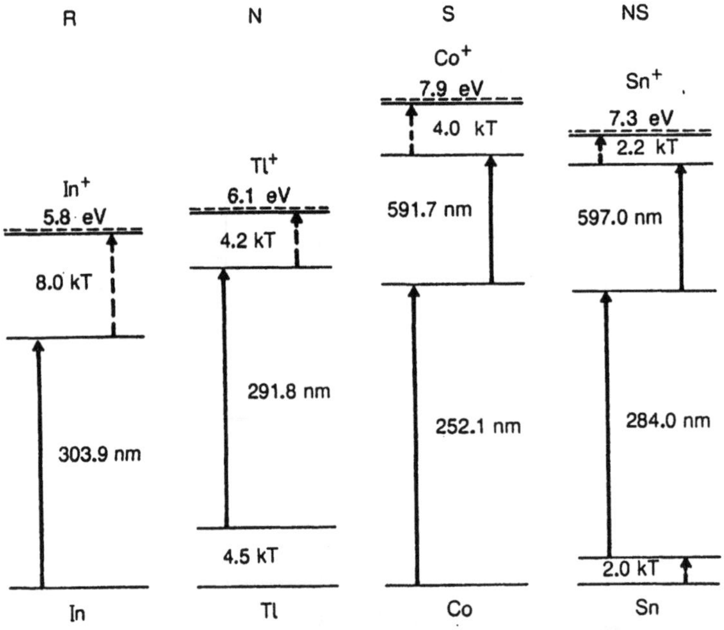

Fig. 3. Photoexcitation schemes for LEI in a 2500 K flame (kT = 1735 cm^{-1}) [2]

dispersion is necessary and the resolution is limited primarily by the laser. The lack of optical detection eliminates interferences from scattered laser radiation, flame background, and ambient (room) light. The compatibility of the analytical flame with LEI spectrometry enhances the utility of this technique because the flame is one of the simplest and most versatile methods of sample introduction currently available. In addition, collisional processes in the flame are an asset rather than a liability for LEI. LEI spectrometry possesses a combination of properties which make it an attractive choice for analytical flame spectrometry.

The periodic table shown in Fig. 2 indicates the elements which have been determined in analytical flames by LEI spectrometry to date. Limits of detection are given in nanograms of analyte per milliliter of distilled water aspirated into the flame. One ng/ml corresponds to to an atom density of approximately $10^8/cm^3$ in the flame [24]. The table also shows whether the element was determined using a single wavelength or stepwise excitation scheme. The possible photoexcitation schemes for LEI are identified in Fig. 3. Fig. 2 reports LEI detection limits which are competitive with — and in many cases superior to — those obtained with other techniques of atomic spectrometry.

2 Signal Production and Collection

2.1 Ion Production

In its simplest form, LEI is a two-step process (see Fig. 3). It involves three quantum states: the atomic ground state, an atomic excited state, and an ionic ground state. For excited levels very near the ionization potential, ionization rates approach collision rates, giving ion yields near unity. The essential steps for LEI, photoexcitation and thermal ionization, are not the only processes occurring in an atmospheric pressure flame. An excited atom can also be collisionally deactivated or fluoresce. A detailed description of signal production requires a complex expression involving several competing rate constants [25].

The probability of ionization of a given atom or molecule on collision is governed by the Arrhenius factor, $\exp[-(E_i - E_j)/kT]$ where E_i is the ionization potential and E_j is the energy level occupied by the atom or molecule, k is the Boltzmann constant, and T the flame temperature. The collisional ionization probability for a low-lying atom may be increased by two orders-of-magnitude by an eV of optical excitation, making LEI a viable approach for sensitive determinations of trace metals.

A helpful qualitative understanding of the dynamics of ion production in LEI may be gained by considering the hydrodynamic analogy illustrated in Fig. 4. The tubs represent the three energy levels, with the liquid levels indicating the atom or ion populations. (State multiplicities are ignored for simplicity, i.e., statistical weights are assumed to be equal.) The pumps representing the laser and thermal energy must have pumping rates proportional to the pressure head (population) as well as the rotational velocity of the pump rotor (laser power/Arrhenius factor) for the analogy to be accurate.

In the absence of laser excitation, the fluid level in the top two tubs is negligible when compared with the bottom tub. Figure 4 illustrates the fluid levels some time

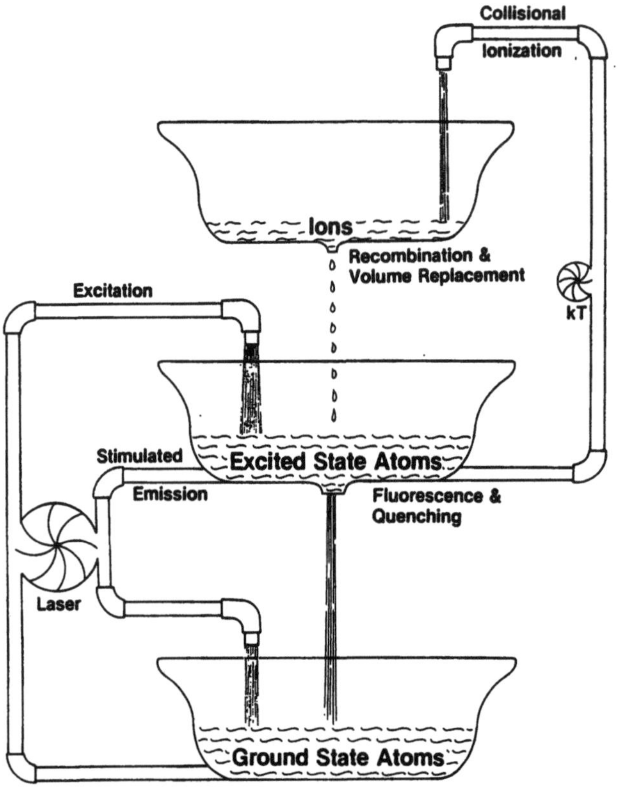

Fig. 4. Hydrodynamic analogy to LEI for resonant photoexcitation scheme (R of Fig. 3)[2]

after the laser pulse begins. The fluid levels (or populations) in the bottom two tubs have equilibrated under optical saturation. Collisional ionization from the excited state relentlessly depletes the neutral population (the bottom two tubs) while generating an ion population. Although in this example, the total system has not yet equilibrated due to the relatively slow rate of ion-electron recombination, ionization will approach completion if the laser remains on long enough. The consequences of laser pulse duration and the ionization rate have been examined leading to the following rule-of-thumb: Unit ionization efficiency may be approached with an optically-saturating laser pulse whose duration significantly exceeds the reciprocal of the effective ionization rate of the laser-populated excited state[2]. The qualifier "effective" in the above statement accomodates the effect of Boltzmann-populated states above the state in question[26–28].

2.2 Ion Collection

Once generated, the collection of analyte ions by applying an electric field is simple. Unfortunately since other ions will also be collected by this scheme, the LEI signal can be influenced by high ion concentrations originating from the sample, the flame, and even the analyte itself. The latter possibility is of little practical consequence because only IA and IIA elements have significant ion fractions in an acetylene/air flame and samples containing these metals as analytes may be diluted if an electrical interference is a problem. Other instances of these electrical interferences with analyte signals have been the subject of many investigations [29–35]. These studies have led to an evolution of the electrode design (Fig. 5) and a better understanding of the signal collection process.

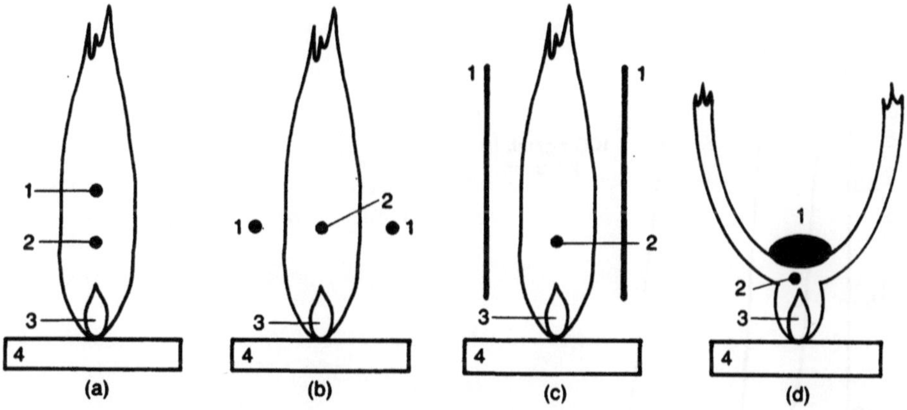

Fig. 5. Evolution of the electrode configurations used for LEI [2]. 1 = high voltage electrode, 2 = laser beam, 3 = reaction zone, 4 = burner head. (a) and (b) 1 is a rod cathode. (c) 1 is a plate cathode. (d) 1 is a water-cooled cathode. The burner head, 4, is the anode in all cases

Early LEI measurements were made with a cathode in the flame (see Fig. 5a). The signal was taken off the burner head which served as the anode. Soon thereafter the cathode was split and the electrodes were placed just outside the flame (Fig. 5b). This configuration was attractive because it was nonintrusive and the tungsten electrodes were not subject to deterioration in the flame. Unfortunately, this configuration led to severe electrical interferences.

Figure 6 illustrates the LEI signal behavior for the external split cathode when the aqueous samples contain different concomitants with low ionization potentials in addition to the indium analyte. The laser was tuned to the 303.9 nm indium line. Although electrical interferences are significant only when the concomitants are IA or some IIA elements, this interference makes the analysis of some real samples more difficult. For this reason, it is worthwhile to understand the cause of the interferences and discuss possible remedies.

Signal suppression has been explained by the formation of a space charge at the cathodes [27–31]. The distribution of charge around a probe extended into a plasma can be divided into three regions [36a]. If the probe is slightly negatively biased, a region will form near the surface of the cathode where the electron concentration is much lower than the ion concentration. The excess positive charge at the cathode is a consequence of the great disparity in the mobilities of ions and electrons. The velocity of electrons is 100–1000 times greater than positive ions at the same field

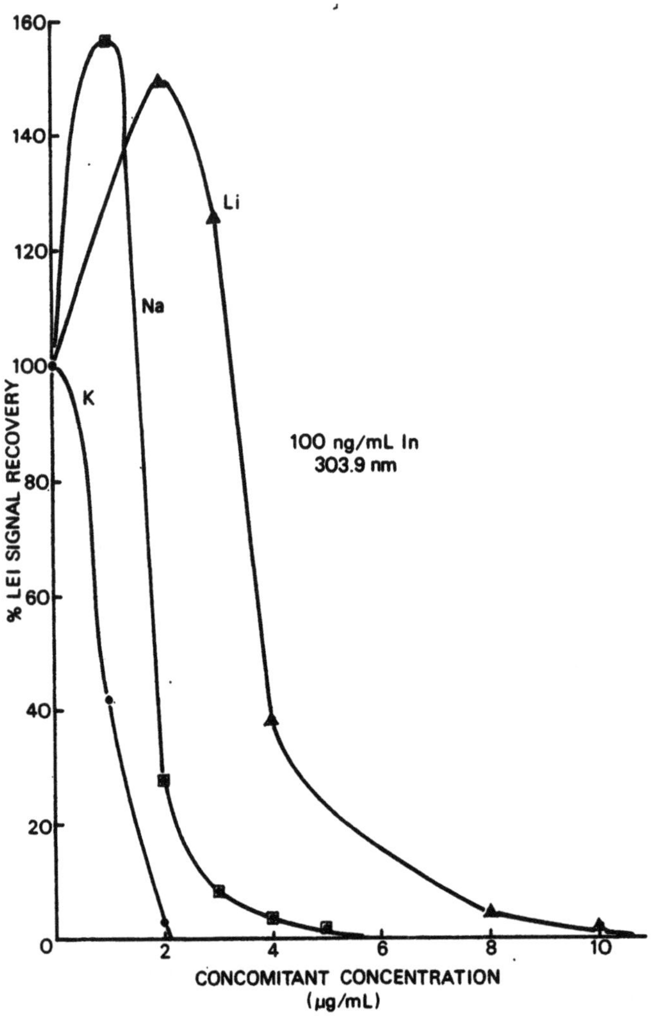

Fig. 6. LEI signal recovery curves for indium in potassium, sodium, and lithium sample matrices

strength [36b]. This region has been referred to as the sheath. Just outside the sheath, a transition region exists where charge separation begins. In the bulk of the plasma, the concentration of cations and electrons is essentially equal. The nonzero field necessary for LEI signal detection exists only in the sheath and the transition region. However, for large diameter or flat cathodes, with on the order of —2000 V applied and typical flame ionization rates, the sheath may extend for 1 cm or more from the cathode.

Figure 7 illustrates the improvement in LEI signal recovery which occurs when the diameter of cylindrical electrodes is increased. The use of planar (plate) cathodes considerably improves the tolerance of the LEI signal to high levels of ionization in the flame [33]. The reduction in electrical interferences is due to the commensurate

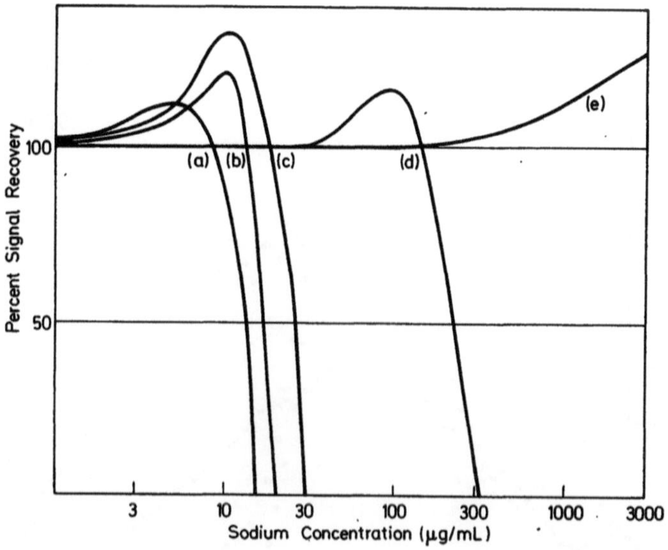

Fig. 7. The effect of electrode design on LEI signal recovery in the presence of varying sodium matrix concentrations: (a) 1-mm diameter rods, (b) 1.5-mm diameter rods, (c) 2.35-mm diameter rods, (d) 5-mm wide × 125-mm long plates, (e) water-cooled immersed cathode [2]

reduction in the field strength at the electrode surface. The high fields near small-diameter rods exacerbate the interference.

To a certain extent, the LEI signal may be recovered by increasing the applied voltage since the sheath expands with increased voltage. This approach is limited because when the applied voltage reaches a certain level, electrical breakdown (arcing) will occur through the flame.

The voltage at which the sheath just extends to the anode is referred to as the *saturation voltage* [36c]. At voltages higher than saturation, a nonzero field fills the region between the cathodes and the anode and every ion and electron produced by thermal ionization will be collected. For voltages above saturation, the *saturation current* is constant. Current vs. voltage curves have been used to evaluate electrode designs and characterize interferences [34, 37].

Maps or images of LEI ions and electrons have been obtained by taking the signal from a small rod positioned between anode and cathode plates [37]. Figure 8 shows the results of this experiment and the electrode configuration used. When the rod is translated vertically it intercepts the ions which are traveling to the electrode as a function of position. At high voltage, the images for electrons and ions are centered at the same height above the burner head as the laser beam. Even at low voltages, although the images are shifted by the contributions of flame velocity and diffusion, essentially all of the signal is collected. Recombination, another possibility for loss of LEI signal, does not occur at a rate which is sufficient to deplete the ion concentration. Therefore, the probability of collecting 100% of the LEI signal is very high.

As seen in Fig. 7, a water-cooled, immersed cathode (Fig. 5d) provides the most resistance to interferences due to high ion concentrations in the flame [34]. This electrode may be constructed by slightly flattening a 0.25 in. diameter stainless steel

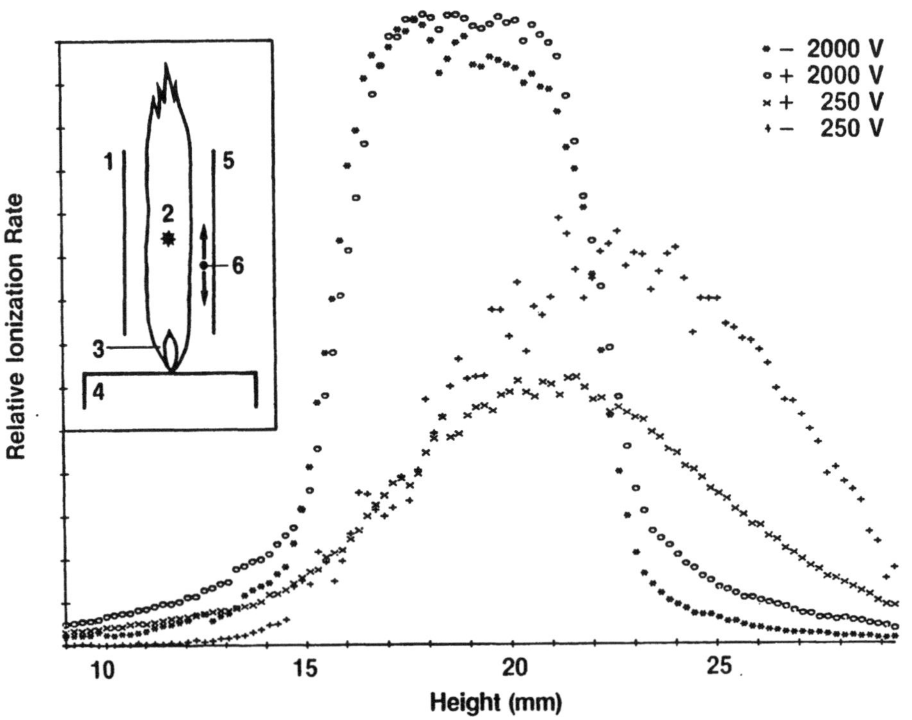

Fig. 8. Images of LEI ions and electrons, obtained by taking the LEI signal from a thin rod translated across the front of the normal collecting plate at the indicated high voltages [49]. The experimental apparatus is shown in the inset: 1 high voltage repelling plate, 2 laser beam, 3 flame reaction zone, 4 burner head, 5 low voltage electrode plate, 6 vertically movable signal pick-off wire

tube. The flat surface, which is generally ground smooth, simulates a plate electrode. Water circulation prevents degradation of the electrode itself. Since the cathode is positioned in the flame, the region close to its surface is a suitable sampling volume. With external cathodes (Figs. 5b and 5c), the electrode surface is not adjacent to the region of maximum atom concentration (and maximum ion production) and the cathodes and the flame are separated by an air gap. With the immersed electrode, ions produced by laser enhancement remain within the collecting field even when the sheath shrinks toward the cathode as a result of high ion concentrations in the flame.

Although it appears that suppression may still be observed with the immersed electrode if the concentration of the concomitant is increased beyond 3000 µg/ml, other considerations due to the high salt concentration of the sample, such as burner clogging and arc-over, may be more important than signal suppression. When extremely high salt concentrations are present, flame analysis by other spectroscopic techniques becomes problematic as well.

Studies in which the voltage applied to the external split cathodes was pulsed have illustrated the formation of the capacitive double-layer which is responsible for signal suppression [35]. The maximum LEI signal could be recovered only if the 1 µs laser pulse was delayed a minimum of 1.5 ms after the initiation of the 4 ms high voltage pulse. Figure 9 shows the results of potential measurements made with a tungsten rod replacing the laser beam while pulsing the applied voltage on the split cathodes. The burner head was used as the anode. These discharge time response curves indicate a time-dependent increase in the LEI signal collected by the external split cathodes. Addition of a low ionization potential concomitant further increased the time constant for field development. When the split cathodes were placed in contact with the flame, the maximum LEI signal was observed after only 200 μs. Using an immersed cathode, centered in the flame, and exciting near the electrode surface permitted the recovery of LEI signals after only 8 μs. Therefore, not only are interferences avoided with the immersed electrode but signal collection is improved because the diffusion path in the flame is reduced. Observation of the LEI signal with laser delays less than 8 μs was not possible because the signal was obscured by a large voltage spike originating from the capacitor which was used to separate the signal from the background.

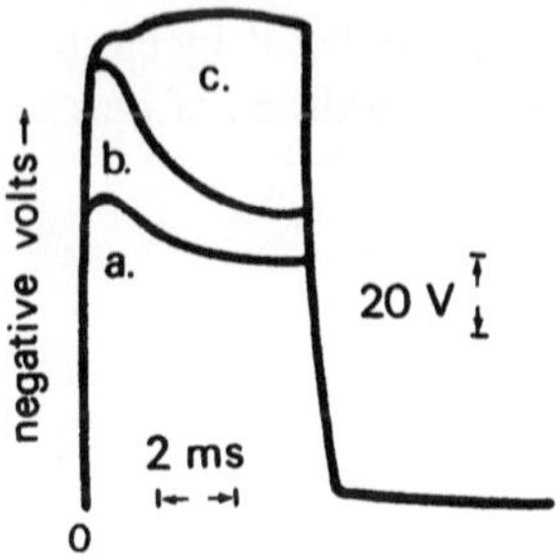

Fig. 9. Discharge time response curves in the flame measured at the laser beam position: **a** 100 μg/ml In, **b** 100 μg/ml In with 1 μg/ml K, **c** 100 μg/ml In with the cathodes moved into the flame [35]

The LEI signal produced by amplitude-modulated *continuous wave* (CW) dye laser excitation has been shown to be less concomitant-dependent than signals obtained with pulsed excitation [38]. CW excitation is almost completely immune to interferences from low ionization potential sample matrices at virtually any position in an acetylene-air flame whereas pulsed excitation produces the maximum signal recovery only near the cathode surface. CW is more tolerant in this regard because convection or diffusion will move the analyte ion into the nonzero field near the cathode surface during the synchronization window for chopping rates less than 500 Hz.

Time resolution of LEI signals has been demonstrated as an approach for discriminating against electrical interferences [35]. When a sample is aspirated into a premixed burner, it is diluted in the mixing chamber prior to introduction into the flame. The sample concentration reaches a steady state value after a relatively short time. The

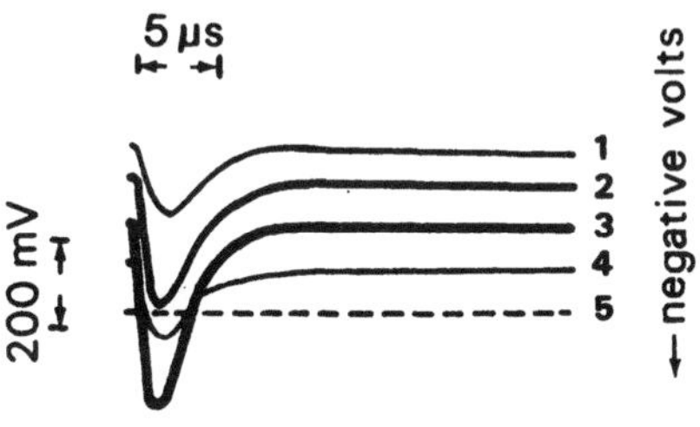

Fig. 10. Oscilloscope traces of sequential LEI signals for 100 μg/ml In with 5 μg/ml K [35]. Note that while the transient signal increased to a maximum (1–3) and then decreased to zero with time (4, 5), the background increased linearly

sequence is reversed when sample aspiration is terminated. Figure 10 illustrates the sequence of events when 100 μg/ml indium was the analyte with 5 μg/ml potassium as the concomitant. The indium signal reached a maximum a few ms after aspiration was initiated and then decreased to the level which would be observed in a steady state experiment, i.e., the indium signal was completely suppressed by the presence of potassium. Recovery of the transient LEI signal for the analyte is the instrumental analogue of sample dilution to reduce concomitant interferences.

Because signal enhancement does not constitute as severe a problem for the analyst as suppression, it has not received as much attention. Although it may be advantageous in some cases, signal enhancement still qualifies as an interference. The use of the immersed electrode which has largely eliminated electrical interferences, at least for acetylene-air flames, does cause signal enhancement when a high concentration of ions is present. Speculation on the cause of LEI signal enhancement suggests that it may be related to the increasing field strength in the sheath as it compresses when a low ionization potential concomitant is added. This effect is predicted by electric field maps [34]. This increase in the field strength produced by the addition of ions to the flame has also been observed experimentally [35].

3 Analytical Considerations

3.1 Instrumentation

A schematic diagram of a typical LEI spectrometer is shown in Fig. 11. The bulk of LEI spectrometry and all of the detection limits reported in Fig. 2 have been accomplished using a premixed burner. In a premixed burner, the aspirated sample is mixed with the fuel and oxidant prior to combustion. A 10-cm slot burner head for acetylene-air has been used predominantly for LEI spectrometry [29–35] but results for a 5-cm

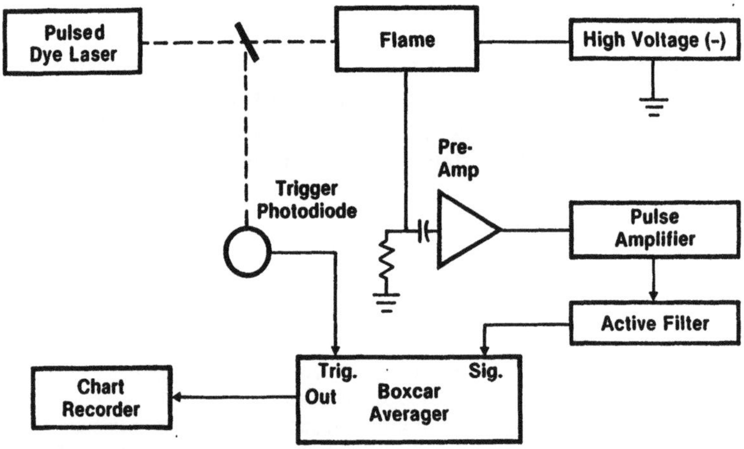

Fig. 11. Block diagram of a typical LEI spectrometer

slot burner head with an acetylene-nitrous oxide flame have also been reported [39] and are included in Fig. 2. The acetylene-air flame is the sample reservoir of choice for a large number of metals except those which form refractory oxides in the flame. The higher temperature nitrous oxide flame decomposes these oxides rendering them accessible to atomic spectrometry. Until the advent of the immersed electrode, the nitrous oxide flame was unusable for analytical determinations because of the high level of electrical noise generated by the flame.

Recently, a total consumption (turbulent flow) burner has been examined for LEI spectrometry [40]. Although excellent limits of detection have been reported for LEI spectrometry with premixed burners, a total consumption burner has several potential advantages. In the total consumption burner, the entire sample is aspirated into the flame. This should result in more analyte atoms in the flame, even considering unvaporized solution, producing a net gain in sensitivity. LEI detection is insensitive to scattered source light and so should not be limited by scattering from unvaporized solution droplets. Since the fuel and oxidant are mixed at the burner nozzle, there is no possibility of explosive flashback regardless of the fuel and oxidant. Rapid-propagating, high temperature, oxygen-based flames will be available for LEI spectrometry with the total consumption burner, permitting more effective utilization of atomic transitions originating in excited states. A total consumption burner with a hydrogen-oxygen or acetylene-oxygen flame presents an alternative for LEI spectrometry of metals which form refractory oxides.

Preliminary experiments with total consumption burners have resulted in the observation of substantial LEI signals in hydrogen-air, acetylene-air, hydrogen-oxygen, and acetylene-oxygen flames. Possible electrode configurations for LEI spectrometry with the total consumption burner have been evaluated using indium as the analyte and an acetylene-air flame. The detection limits obtained for indium using the optimum electrode configuration were comparable to the values determined for a premixed burner with a 10-cm burner head. For elements, such as indium, where extremely low LEI detection limits have been obtained, further improvements in sensitivity may be limited by the analyte atom fraction which is determined by the fuel-oxidant combination. The detection limits for both copper and manganese have been lowered using the total consumption burner, the latter by a factor of 3 better than published results with the premixed burner. When the pathlength difference between the 2.5-cm diameter total consumption burner and the 10-cm premixed burner is considered, the results obtained with the former are even more impressive. The detection limits obtained with the total consumption burner may be improved by desolvation of the sample, confinement of the sample to a smaller volume, and pathlength extension.

LEI spectrometry using the total consumption burner, with greater sample throughput and a wider range of usable fuel-oxidant combinations, expands the possibilities for development of a more sensitive and versatile detection system for atomic spectrometry. In addition to furthering the analytical methodology, these results demonstrate that high-sensitivity LEI measurements are possible in adverse sample environments where traditional methods of optical spectrometry have proven inadequate.

Many electrode configurations have been used with comparable success for determining the concentration of metals in pure, aqueous solutions. The water-cooled, immersed electrode is preferred for its resistance to electrical interferences, as described in the previous section. The specific configurations used for premixed and total consumption burners differ because of the differences in the maximum analyte concentration zones in the respective flames [40]. A dc voltage applied to the electrodes is currently the best mode of signal collection.

A pulsed dye laser is the most practical excitation source for LEI spectrometry. Amplitude-modulated CW dye lasers have displayed a useful immunity to electrical interferences at any excitation position [38], but because of the inefficiency of the frequency doubling process, CW dye lasers yield low power emission in the ultraviolet spectral region. Most metals have their strongest resonance lines in the ultraviolet. In addition, ultraviolet transitions terminate nearer the ionization limit, producing the highest LEI sensitivity.

As implied in previous sections, a flashlamp-pumped dye laser should be the best source for LEI because of its relatively long (approximately 1 μs) pulse duration. LEI detection limits obtained with laser-pumped dye lasers suffer somewhat because of their reduced pulsewidths (6–10 ns) which may be partially compensated for by their high peak powers. The efficiency of nonlinear second harmonic generation is also improved by higher peak pulse powers. Experience has shown that Nd:YAG laser-pumped dye lasers are capable of producing single wavelength detection limits which are comparable to the best achieved with a flashlamp-pumped dye lasers in many cases. Nitrogen laser-pumped dye lasers perform less well not only because of their lower peak powers but also because of the radiofrequency (RF) interference which is broadcast by the nitrogen laser. When stepwise excitation is required, dye lasers simultaneously pumped by either a nitrogen laser [18] or Nd:YAG [19] laser have been used. Again, the Nd:YAG system is preferred because of its superior peak pulse power, more complete wavelength coverage with the dye fundamental (using Nd:YAG second and third harmonic pumping), and low levels of RF. Using the appropriate optics, one or both dye lasers may be doubled for tunable ultraviolet operation. In addition, when a single laser is used to pump two dye lasers simultaneously, timing synchronization concerns are minimized.

The laser-related current pulses detected with the electrodes may be simply monitored as a voltage drop across a resistor. More commonly, a high-gain preamplifier is used for analytical LEI [41]. An RC network blocks the dc background current from the flame and passes the signal to the processing electronics. To achieve the optimum performance from the preamplifier, it should be mounted as close as possible to the anode where the signal is taken. This minimizes formation of ground loops, reduces cable capacitance, and permits the use of a short, shielded cable to avoid RF interference pickup. Thorough grounding and shielding of the burner system are also effective in reducing environmental noise. This preamplifier will yield approximately 2 μs pulses regardless of the laser pulsewidth because of intrinsic bandwidth limitations. This preamplifier is therefore conveniently resistant to 10–100 ns changes in electron transit times due to matrix and geometry effects [2].

A boxcar averager in the single-point mode has been used for processing the LEI signal [32,33]. The variable-width electronic gate is positioned over the LEI pulse. A trigger signal from a photodiode which monitors the laser beam opens the gate for a specified time after the laser pulse. The signal recovered on successive laser pulses is averaged depending on the instrumental time constant. The average signal amplitude may be read out on a strip chart recorder or converted into a digital signal for further processing with a microcomputer [42]. Generally the signal pulse is displayed on an oscilloscope simultaneously.

3.2 Methodology

A brief discussion of current methodology drawn from the cumulative research should be helpful in crystallizing the best approach to practical LEI spectrometry. At this point, one would choose a premixed burner with a water-cooled, immersed electrode for the reasons discussed in previous sections. When using a premixed burner, it is important to acidify all blank, standard and sample solutions [43]. This procedure prevents adsorption of solutions on the interior of the burner premixing chamber and contributes to larger LEI signals. Otherwise very large transient LEI signals result when the aspiration of an acid solution is alternated with a neutral solution containing the analyte. LEI is effected by sample adsorption in the premixed burner to a much greater extent than other flame techniques because of its greater sensitivity. Addition of nitric acid is preferred because it provides constant signal enhancement over a relatively wide concentration range, making the amount added to samples less critical [43].

A flashlamp-, nitrogen laser-, or Nd:YAG laser-pumped dye laser system may be used for LEI, keeping in mind the tradeoffs. Excimer laser-pumped dye laser systems should also be suitable. Realistically, availability may be the most compelling factor in choosing a laser.

The selection of the best analytical line has not been discussed explicitly and deserves attention. Because LEI is at least a two-step process involving laser excitation and thermal ionization steps, many transitions may be preferred for LEI which are not usable by purely optical methods. In other words, excited states which have low fractional populations may produce good LEI sensitivity due to proximity of the laser-populated state to the ionization potential. This makes the choice of the most sensitive LEI lines more complicated but it introduces an important practical advantage for dye laser spectrometry.

The efficiency of an organic dye for laser action over a specific wavelength range is characterized by its gain curve. The visible (fundamental) and ultraviolet (second harmonic) spectral regions are covered by a series of dyes, but the power output is variable with wavelength. Since more analytical lines are available which can provide good sensitivity using LEI spectrometry, the analyst may choose a line in a spectral region where the dye laser performs most efficiently. By choosing the optimum combination of LEI line and dye, superior sensitivity may be achieved. In addition, the high resolution and tunability provided by a dye laser coupled with LEI detection, will permit the determination of many metals within the gain curve of a single laser dye.

The overlap of molecular spectra with atomic lines, which occurs in optical flame spectrometry, has been less commonly encountered with LEI. Native flame species such as OH and CH are not observed because of their high ionization potentials. Resonantly-enhanced multiphoton ionization of molecules such as NO [45,46] may cause interferences in some flames. The LEI spectra of oxides of lanthanum, scandium, and yttrium have been observed in hydrogen-air flames [47] and may constitute a spectral interference for samples containing these metals.

The larger number of analytical wavelengths available with LEI spectrometry is an advantage for avoiding infrequent atomic and molecular spectral overlaps. The analyte signal can be identified by scanning the dye laser wavelength across the analytical line. Stepwise excitation may improve the selectivity of LEI spectrometry [19]. The probability of both analyte transitions coinciding with two interferent lines is very small. A spectral background may overlap either analytical line but the LEI signal will be enhanced only by the interaction at both wavelengths. Either of the two wavelengths may be scanned while holding the other fixed to distinguish the signal from background. Generally, scanning the visible (fundamental) wavelength is preferred because of the difficulty of tracking a frequency-doubling crystal while tuning the dye laser. A complete spectral overlap of the analyte line with another atom at one wavelength can also be treated by scanning the second laser and keeping the first laser wavelength fixed [19]. An increased baseline signal due to the first step excitation of the interferent is observed but 100% of the analyte signal can still be recovered. Dynamic ranges for LEI have not been studied in detail but ranges of four to five orders-of-magnitude are typical [48].

3.3 Limits to Sensitivity

The limits to sensitivity in LEI spectrometry have been discussed recently [49]. The ultimate limits of detection obtainable are determined by noise associated with the LEI signal. Several sources of noise have been identified [50], but the limiting noise for optimum conditions is statistical or shot noise. A reasonable ultimate detection limit for LEI may be calculated as 1 part-per-trillion (1 pg/ml) or 10^5 atoms/cm^3 [49]. According to Fig. 2, lithium may be detected at this level. Variations in detection limits for other metals may be attributed to incomplete atomization in a given flame and lack of 100% ionization. Of course, when real samples are analyzed, other complications which have already been discussed may arise leading to higher detection limits than obtained for pure, aqueous analyte solutions.

4 Applications

4.1 General

The general application of LEI spectrometry to the determination of trace metals has been somewhat limited but progress in this area should continue. Alloy analyses are particularly amenable to LEI spectrometry because of the absence of an ionizable sample matrix. Indium (303.9 nm) has been determined in a nickel-based high temperature alloy [31]. Atomic absorption spectrometry of this sample requires time-consuming extraction procedures to remove concomitant metals which contribute to spectral interferences. After dissolution of the alloy sample with acids, the concentration of indium was determined to be 35 μg/g by LEI spectrometry.

This agreed within experimental error with a value of 37 μg/g obtained with a graphite furnace. Manganese was also determined in a National Bureau of Standards (NBS) Standard Reference Material (SRM) No. 1261 steel sample to within 0.66% of the certified value [31].

Several low alloy steels (SRM 362 and 363) and unalloyed copper (SRM 396) have been analyzed by stepwise LEI spectrometry [19]. Calibration was performed by bracketing the dissolved SRM samples with standards without matrix matching. The LEI results agreed within experimental error with the certified values in all cases. The stepwise LEI signal was obtained by scanning the dye laser over the second step wavelength with the first step wavelength fixed. The LEI value for tin in the unalloyed copper sample was reported with a factor of 6 less uncertainty than the NBS SRM value. The low detection limit for tin and the absence of any significant spectral background made a high precision measurement possible by LEI spectrometry.

Rubidium has been determined in pine needles (SRM 1575) to demonstrate the resistance to interferences of the immersed electrode with CW laser excitation [38]. The concentration of matrix components such as calcium and potassium were on the order of 40 μg/ml. The measured rubidium concentration of 11.1 μg/g agreed well with the certified value.

4.2 Flame Diagnostics

The application of LEI spectrometry to flame diagnostics has been recently reviewed [51]. This is a research area where the unique aspects of the technique can provide some new insights into combustion systems.

Vertical spatial profiles of a hydrogen-air flame have been generated by interposing a small diameter rod between the anode and cathode plates as illustrated in Fig. 8 [37]. An amplitude-modulated CW laser was the excitation source for LEI. In addition to obtaining ion images, this experiment also demonstrated the physical size of the excess ion region produced by laser enhancement and the influence of external voltage, flame velocity, diffusion, and electrostatic expansion on the excess ion region. The rise and decay times of the excess ion current which were followed at different applied voltages illustrate the dependence of LEI on ion mobility, electric field, and excited-state ionization rate constant. The concepts demonstrated here have contributed to a better understanding of the LEI process but have also indicated the promise of LEI spectrometry for combustion diagnostics.

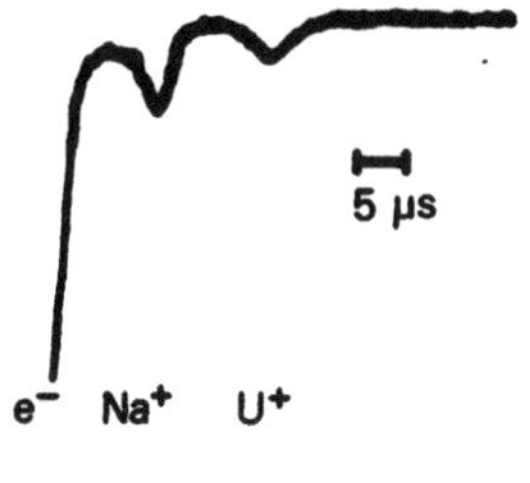

Fig. 12. Oscilloscope trace for irradiation of an acetylene-air flame containing Na and U atoms with a single laser pulse [52]. At the excitation wavelength, 539.9 nm, there is a discrete two-photon sodium transition and a broad band uranium transition

Atomic ion mobilities in several flames have been measured by detecting the arrival time of the ions produced by laser enhancement [52]. The time from the production to the collection of the ions was measured as a function of the distance from the laser beam to the collecting electrode for a series of applied voltages. An oscilloscope trace for irradiation of an acetylene-air flame containing sodium and uranium with a single laser pulse is shown in Fig. 12. The arrival of the electron pulse and the ion signals due to sodium and uranium are visible. The mobility measurements for lithium, sodium, potassium, calcium, iron, strontium, barium, indium, thalium, and uranium were compared to the Langevin theory. The model predicted the mobility of some of these ions to within 10%, provided that the dielectric constant of the flame gases was considered. In general, the larger, more polarizable ions were subject to the largest deviation, sometimes as much as 50% larger than predicted. For atomic ions, the Langevin theory was accurate and provided an upper limit for estimating absolute ion mobility.

Using the same apparatus, the mobility of very small particles at the sooting limit of a premixed flame has been estimated [53]. The observed ionization signals were attributed to laser-induced heating and subsequent thermal ionization of the soot particles. An approximate estimate of the particle size was made from the mobility measurements.

The temporal and spatial evolution of the depleted neutral atom density following LEI has been used to characterize the flow velocity of the flame gases in a laminar flow flame [54]. A CW dye laser beam was split into two beams and tuned to the 589.0 nm sodium transition. The lower beam was acousto-optically modulated and produced the depleted sodium atom region in the flame. The upper beam (probe) was directed counter to the modulated beam through a Na D line filter onto a photomultiplier tube. The LEI laser was operated in a pulsed mode and the output from the photomultiplier which monitored the sodium absorption was sent to a signal averager. The averaged signal was processed by a small computer which fit the signal to a model for arrival time. This measurement was repeated for two probe beam heights. The differences in the arrival times and the probe heights yielded the velocity at the probe's average height above the burner with 2% precision. This technique is limited to reasonably laminar flames because turbulence will dilute the neutral atom depletion region. No flow perturbations are introduced by optical detection of the LEI.

5 The Future of LEI

Research must be undertaken to demonstrate that LEI is adaptable to a wider variety of samples and analytically-useful flames. This will require further consideration of methods to discriminate against or remove low ionization potential interferents. Preliminary results have indicated that the use of an acetylene-nitrous oxide flame for the determination of metals which form refractory oxides exacerbates electrical interferences when samples contain IA elements [39]. The much higher flame temperature produces higher concentrations of ions whose effects cannot be entirely mitigated by using an immersed electrode.

Prior removal of interfering ions with an additional set of biased collection electrodes [55] or chromatography are possibilities. The latter approach has been successful for the determination of trace metals in seawater by graphite furnace/atomic absorption spectrometry which also suffers from high salt interferences [56]. Pulsing the applied voltage to an immersed cathode could discriminate against electrical interferences if some practical signal processing problems can by overcome [35]. CW laser excitation may circumvent some of the signal collection interferences [38]. Currently, the limited wavelength coverage of commercial CW dye lasers is not very attractive for analytical atomic spectrometry. Higher efficiency ultraviolet generation is favored by the high peak power available with pulsed lasers. Improvements in laser technology could alleviate this situation in the future.

Although it possesses many advantages in terms of versatility, ease of operation, and low cost, the flame has its limitations for high sensitivity analytical spectrometry. Further lowering of detection limits will require other sample reservoirs. For example, atomization in a plasma could be coupled to a low temperature, low background flame for laser excitation. Electrical interferences might also be eliminated using such a segregated approach.

LEI spectrometry has been demonstrated as a highly sensitive and selective method for analytical determinations in flames. Since an electrical signal is measured directly, LEI detection is insensitive to scattered excitation, flame background and ambient light. This property accounts for the very low detection limits which have been obtained for metals in a turbulent flow, total consumption flame. These results suggest that LEI spectrometry may be useful for sensitive measurements in adverse sampling environments where conventional optical methods have been unsuccessful.

Laser-enhanced ionization spectrometry should occupy a prominent position in analytical methodology because of the high sensitivity and precision which it provides. Because of the simplicity of the detection scheme, LEI is an easily-implemented and useful complement to other laser-based techniques as well.

6 Acknowledgments

The author wishes to acknowledge the National Science Foundation for partial support of this research. The many people who contributed to the body of work described here are represented in the references. J. C. Travis and J. E. Hall made many valuable suggestions during the preparation of this manuscript. The author would also like to thank P. K. Schenck and G. C. Turk for their help in preparing Fig. 2.

7 References

1. Green, R. B. et al.: J. Am. Chem. Soc. *98*, 1517 (1976)
2. Travis, J. C. et al.: Anal. Chem. *54*, 1006A (1982)
3. Young, J. P. et al.: ibid. *51*, 1050A (1979)
4. Hurst, G. S. et al.: Rev. Mod. Phys. *51*, 767 (1979)
5. Hurst, G. S.: Anal. Chem. *53*, 1448A (1983)
6. McClain, W. M.: Accts. Chem. Res. *7*, 129 (1974)
7. Kramer, S. D. et al.: Opt. Lett. *3*, 16 (1978)
8. Whitaker, T. J. et al.: Chem. Phys. Lett. *79*, 506 (1981)

9. Mayo, S. et al.: Anal. Chem. *54*, 553 (1982)
10. Fassett, J. D. et al.: ibid. *55*, 765 (1983)
11. Balykin, V. I. et al.: JETP Lett. *26*, 357 (1977)
12. Gelbwachs, J. A. et al.: IEEE J. Quantum Electron, *QE-14*, 121 (1978)
13. Wright, J. C.: Applications of lasers in analytical chemistry, in: Applications of Lasers to Chemical Problems (ed. Evans, T. R.): p. 105, New York, John Wiley & Sons 1982
14. van Dijk, C. A. et al.: Anal. Chem. *53*, 1275 (1981)
15. Lin, K. C. et al.: Chem. Phys. Lett. *90*, 111 (1982)
16. Goldsmith, J. E. M.: Opt. Lett. *7*, 437 (1982)
17. Goldsmith, J. E. M.: J. Chem. Phys. *78*, 1610 (1983)
18. Turk, G. C. et al.: Anal. Chem. *51*, 2408 (1979)
19. Turk, G. C. et al.: ibid. *54*, 643 (1982)
20. Gonchakov, A. S. et al.: Anal. Lett. *12*, 1037 (1979)
21. Zorov, N. B. et al.: J. Anal. Chem. *35*, 1108 (1980)
22. Popescu, D. et al.: Phys. Rev. *A9*, 1182 (1974)
23. Johnson, P. M. et al.: J. Chem. Phys. *62*, 2500 (1975)
24. Winefordner, J. D. et al.: Anal. Chem. *36*, 1939 (1964)
25. Travis, J. C. et al.: ibid. *51*, 1516 (1979)
26. Hollander, T. J.: AIAA J. *6*, 385 (1968)
27. Smyth, K. C. et al.: What really does happen to electronically excited atoms in flames? in: Laser Probes for Combustion Chemistry (ed. Crosley, D. R.), ACS Symp. Ser. *134*, p. 175, Washington, D.C., Amer. Chem. Soc. 1980
28. van Dijk, C. A. et al.: Combust. Flame *38*, 37 (1980)
29. Turk, G. C. et al.: Anal. Chem. *50*, 817 (1978)
30. Travis, J. C. et al.: ACS Symp. Ser., *85*, 91 (1978)
31. Turk, G. C. et al.: Anal. Chem. *51*, 1890 (1979)
32. Green, R. B. et al.: Appl. Spectrosc. *34*, 561 (1980)
33. Havrilla, G. J. et al.: Anal. Chem. *52*, 2376 (1980)
34. Turk, G. C.: ibid. *53*, 1187 (1981)
35. Nippoldt, M. A. et al.: ibid. *55*, 554 (1983)
36. Lawton, J. et al.: Electrical Aspects of Combustion, a) p. 167, b) p. 320, c) p. 315, Oxford, Clarendon Press 1969
37. Schenck, P. K. et al.: J. Phys. Chem. *85*, 2547 (1981)
38. Havrilla, G. J. et al.: Anal. Chem. *54*, 2566 (1982)
39. Peters, R. A.; Green, R. B.: Abstracts, 185th Nat. Amer. Ähem. Soc. Meeting, Seattle, WA., March 24, 1983, No. 209
40. Hall, J. E. et al.: Anal. Chem. *55*, 1811 (1983)
41. Havrilla, G. J. et al.: Chem. Biomed. Environ. Instrum. *11*, 273 (1981)
42. Vickers, G. H. et al.: ibid. *12*, 289 (1983)
43. Trask, T. O. et al.: Anal. Chem. *53*, 320 (1981)
44. Mallard, W. G. et al.: J. Chem. Phys. *76*, 3483 (1982)
45. Rockney, B. H. et al.: Chem. Phys. Lett. *87*, 141 (1982)
46. Smyth, K. C. et al.: J. Chem. Phys. *77*, 1779 (1982)
47. Schenck, P. K. et al.: ibid. *69*, 5147 (1983)
48. Travis, J. C. et al.: The optogalvanic effect, in: Lasers in Chemical Analysis (ed. Hieftje, G. M. et al.), p. 93, Clifton, N.J., Humana Press 1981
49. Travis, J. C.: J. Chem. Educ. *59*, 909 (1983)
50. Turk, G. C.: Ph. D. Dissertation, Univers. of Maryland, 1978
51. Schenck, P. K. et al.: Opt. Engineer. *20*, 522 (1981)
52. Mallard, W. G. et al.: Combust. Flame *44*, 61 (1982)
53. Smyth, K. C. et al.: Combust. Sci. Tech. *26*, 35 (1981)
54. Schenck, P. K. et al.: Appl. Spectrosc. *36*, 168 (1982)
55. Trask, T. O. et al.: Spectrochim. Acta, *38 B*, 503 (1983)
56. Kingston, H. M. et al.: Anal. Chem. *50*, 2064 (1978)
57. Chaplygin, V. I. et al.: Spectrochim. Acta, *38B*, Supplement, 386 (1983)
58. Turk, G. C. et al.: Laser-Enhanced Ionization Spectrometry for Trace Metal Analysis, in: Proc. Colloque Internat. CNRS No. 352, Editions de Physique, Aussois, France, June 20–25, 1983

Laser Spectroscopy of Biomolecules

Angelika Anders

Institut für Biophysik, Universität Hannover, D-3000 Hannover 1, FRG

Table of Contents

1 Introduction 24

2 Biomolecules — General Aspects 24

3 Methods and Instruments 28

4 Investigations of Biomolecules 29
4.1 Chlorophyll 29
4.2 Nucleic Acids and Nucleic-Acid-Dye-Complexes 32
4.2.1 Nucleic Acids 32
4.2.2 DNA-dye-Complexes 35
4.2.3 General Remarks 36
4.3 Hemoglobin, Myoglobin 37
4.4 Rhodopsin 39
4.5 Miscellaneous Topics 40

5 Selective Excitation 41

6 Photomedicine 45

7 Conclusion 47

8 References 47

Spectroscopic applications of lasers on biomolecules such as nucleic acids, chlorophyll, hemoglobin and rhodopsin are reviewed. General aspects, typical examples, applications in photomedicine as well as future possibilities are presented.

1 Introduction

Lasers became indispensable tools in many fields of spectroscopy because of their great advantages over conventional light sources regarding spectral, spatial and time resolution [1,2,3]. They gained growing importance not only in physics and chemistry [4,5,6] but in biology and medicine, too [7,8,9,10]. Laser investigations increase our knowledge about structures and interactions of biomolecules as well as about kinetics of various biochemical processes.

The many possibilities for using lasers in the investigation of biomolecules depend on their various properties. Table 1 summarizes the properties of lasers with typical examples of their application.

Table 1. Laser properties (left-hand side) and typical applications (right-hand side)

high intensity and monochromaticity (i.e. high spectral intensity) tunability	spectroscopy of biomolecules spectroscopy of biological material narrowband irradiation in phototherapy and photochemotherapy selective photobiological effects action spectra
short pulses	time resolved spectroscopy fast reactions
small focus	irradiation of selected parts in cells or tissues
coherence	Doppler spectroscopy, correlation spectroscopy

High intensity and monochromaticity, resulting in a high spectral intensity, are ideal tools for spectroscopic investigations, especially for fluorescence measurements with low quantum yields, for the study of multi-photon processes and excited states, and for Raman spectroscopy. For example, important biomolecules like nucleic acids have an extremely low fluorescence quantum yield at room temperature.

An application in photomedical research is the measurement of absorption and transmission of thick specimens e.g. human skin. A spectral narrowband irradiation of skin lesions and tumors can use the high spectral intensity together with the tunability of special lasers. Action spectra of phototherapeutic interest such as of photosensitizers like psoralens can be investigated. Furthermore, tunable lasers perform a selective excitation of practically any quantum state of atoms or molecules in the wavelength range from about 200 nm to 20 μ.

Short pulses give the possibility of observing the lifetimes and very fast reactions of biomolecules on the picosecond time scale. The small focus of about 1 μ permits exposure of very small selected sites within cells or tissues.

2 Biomolecules — General Aspects

Among the open problems in photobiology which may be investigated with lasers are: UV and visible radiation effects on cells, photosynthesis, vision, photomovement; photomorphogenesis, photo nitrogen fixation, photoactivation of enzymes, photoantagonism, photosynergism, bioluminescence, photosensitization of cells and photomedical questions [11,12]. The receptor molecules e.g. of photosynthesis (chlorophyll and accessory pigments), of vision in vertebrates (rhodopsin and iodopsin) and of photomorphogenesis (phytochrome) are known. The part of a cell most sensitive to UV-light is the DNA (desoxyribonucleic acid) which carries the genetic information.

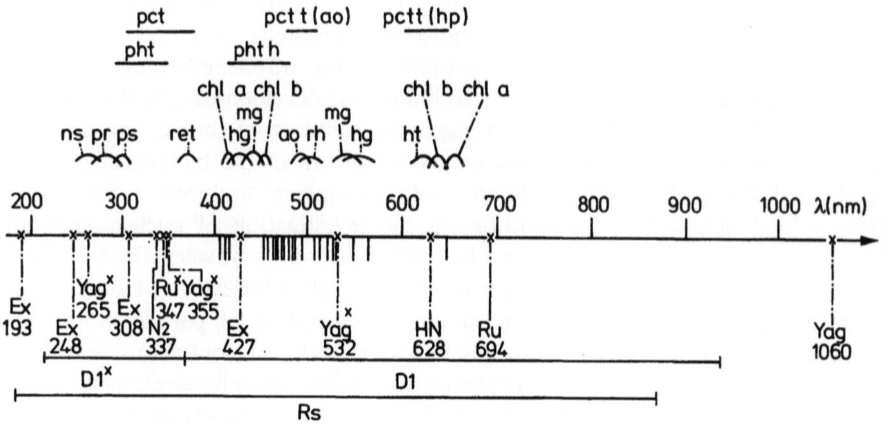

Fig. 1. Main laser lines resp. wavelength ranges (lower part) and absorption ranges of some characteristic biomolecules resp. fields of action in photomedicine (upper part) in the visible and UV wavelength region. As a survey the spectra are shown in the ranges of their main maxima only (see also text).

Ex: excimer laser
YAG: Nd:YAG laser
N_2: nitrogen laser
HN: helium neon laser
Ru: ruby laser
D1: dye laser (different pumping sources)
x: frequency doubled or frequency mixed
Rs: stimulated Raman scattering in hydrogen gas (excitation source: Nd:YAG pumped dye laser)
l: different argon and krypton laser lines
ns: nucleic acids
pr: protein
ps: psoralens
ret: 11.-cis retinal
rh: rhodopsin
mg: myoglobin
hg: hemoglobin
chl: chlorophyll
ao: acridine orange
ht: hematoporphyrin
pht: phototherapy of various dermatoses
pht h: phototherapy of hyperbilirubinemia
pct: photochemotherapy of various dermatoses
pct t: photochemotherapy of tumors

Figure 1 gives a survey about the most important types of lasers required for spectroscopic use and the absorption ranges of some characteristic biomolecules and fields of action in medicine in dependence upon the wavelength in the visible and UV-light.

Biomolecules like chlorophyll, rhodopsin and hemoglobin can be investigated by visible excitation because of their visible absorption bands. But other biomolecules such as nucleic acids and proteins, absorb only in the UV region (see Fig. 2, too). Action ranges of phototherapy and photochemotherapy also lie in the UV range (290–370 nm) (Chapter 6).

The possibilities using lasers in photobiology and in photomedicine depend on the development of laser technology, particularly on whether the laser properties required (such as intensity, bandwidth, pulse or continuous wave operation, pulse length and pulse repetition rate) are available in the wavelength ranges needed. Partly because of their generally better technological development, fixed frequency lasers were often applied. The use of a fixed wavelength is, of course, only possible when it coincids with the absorption or action spectra of the biomolecules to be investigated. This is, for example, the case with a frequency doubled Nd:YAG-laser (530 nm) and hemoglobin. In addition, in the investigation of nucleic acids an Nd: YAG-laser (fourth harmonic) can be applied; having a wavelength of 265 nm, it lies within the absorption maximum of nucleic acids. However, in biomolecules the dependence of certain processes on the wavelength is of particular interest. Therefore, the use of tunable lasers (especially pulsed and continuous wave dye lasers) for investigating biomolecular processes shall be especially mentioned.

For the discussion of spectroscopic problems in photobiology that impact on photomedicine, the wavelength region near 300 nm is of special interest (Fig. 2). The long wavelength absorption edges of the nucleic acids are situated in this wavelength range. Near 300 nm also the maximum of the erythema action spectrum (action spectrum of sunburn) is to be found. At the same time, this is the region where mutations in DNA are induced, under natural conditions from sunlight on the earth's surface, because shorter wavelengths are absorbed in the ozon layer of the atmosphere.

Figures 3 and 4 survey the wavelength ranges and time scales of molecular processes so far as they concern photobiological questions. Furthermore, the regions attainable

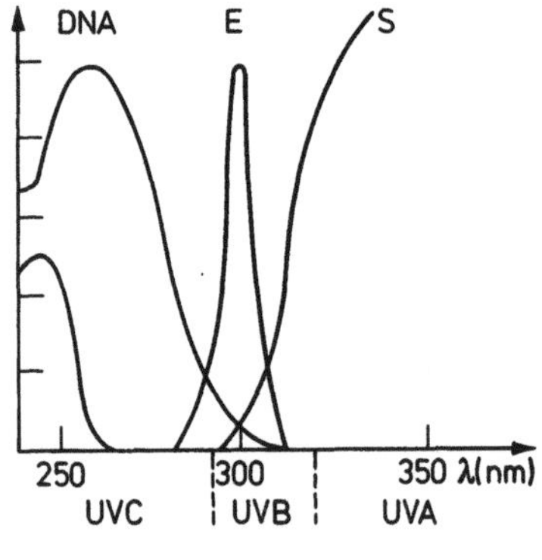

Fig. 2. DNA absorption, erythema effectiveness curve (E) and typical terrestrial solar spectrum (S). UVA and UVB are wavelength regions in which phototherapy and photochemotherapy are applied

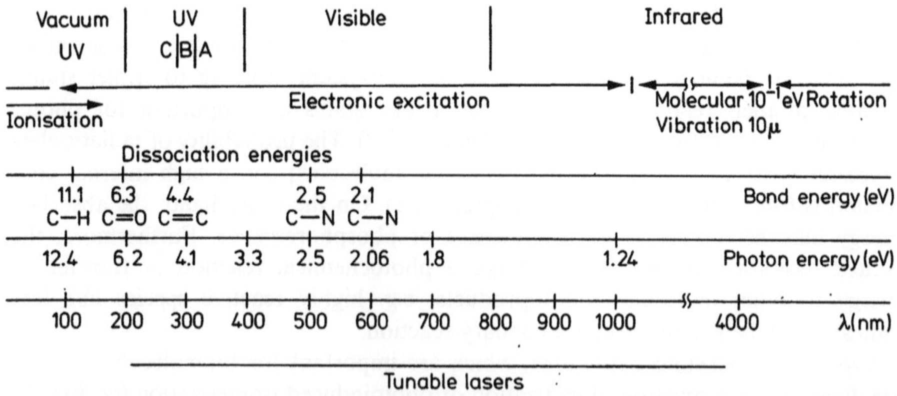

Fig. 3. Ultraviolet, visible and infrared regions of the electromagnetic spectrum with indication of photophysical processes and wavelength regions of tunable lasers

with lasers are indicated. Wavelengths between 200–300 nm inactivate, whereas wavelengths between 300–700 nm may activate and reactivate biological systems. The biological response after light exposure of cells takes place from milliseconds to years (e.g. sunlight-induced human skin cancer).

Visible light is absorbed by pigmented cells and even slightly by unpigmented cells when cytochrome and flavoproteins are present [11,12]. In the presence of pigments (e.g. chlorophyll; see Fig. 1) in the cell, the pigments serve as primary receptor molecules for a photobiological reaction. The effects of visible light are strongly increased by the addition of dyes complexing with biomolecules and acting as photosensitizers in the cell (see Chapter 6). UV-light — as mentioned above — is absorbed by nucleic acids and proteins (Figs. 1 and 2). Furthermore, UV-light may cause mutations, inhibition of cell mitosis, DNA-, RNA- and protein-synthesis as well as sunburn and skin cancer in man. Most cells have developed systems (e.g. photoreactivation) enabling them to repair such UV-light induced lesions [11]. Infrared light is absorbed by water, and cells are composed to about 80% of water. But the energy absorbed from natural light is not enough to raise the temperature in the cells.

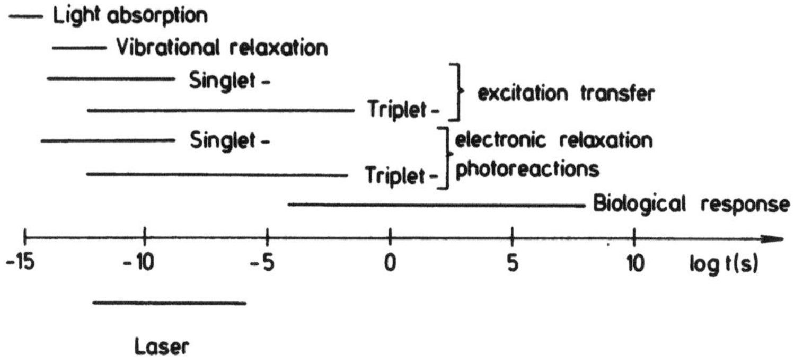

Fig. 4. Time scales of photophysical and photobiological processes with the range of pulsed lasers

After light absorption and electronic excitation of molecular singlet states [e.g. [2,13]] the following deactivation processes may occur: radiationless relaxation by internal conversion to the ground state or intersystem crossing to triplet states. Because of their relatively long lifetimes triplet states are important for photochemical reactions (for psoralens, see Chapter 4.2.2). The probability of radiationless processes is strongly enhanced in polyatomic molecules; e.g. a high internal conversion is found in nucleic acids (Chapter 4.2.1). On the other hand, the absorbed energy may be dissipated by fluorescence or phosphorescence. Furthermore, the excited receptor molecule may undergo a photochemical reaction or transfer its energy to other molecules. Photoproducts, e.g. highly reactive species like free radicals, may be produced in a secondary reaction.

Typical photochemical reactions, which are important for biomolecules, too [14], are dissociation, ionization, dimerization or photoinduced isomerization (cf. Fig. 3). Thus, photodissociation of the sulphur bond has been observed in the amino acid cystine to be most likely the dominating photochemical reaction of proteins excited by UV-radiation. As the cystine-absorption is rather low in the UV, intramolecular energy transfer processes from other amino acids must be important for this reaction. The formation of a dimer between two adjacent thymine bases on the same DNA strand constitutes the predominant biological effect of UV-radiation (Chapter 4.2). Having its structure changed in such a way, the DNA looses its ability to transscribe the information at this site. Intramolecular energy transfer processes along the DNA to thymine having the lowest triplet state of all bases play an important role in this reaction.

An essential naturally ocurring photoinduced isomerization is the cis-trans reaction associated with the primary reaction in vision (Chapter 4.4). Photoionization is only rarely observed, because the ionization energies are normally to high to be supplied by a single UV-photon (see Chapter 5).

3 Methods and Instruments

Dye lasers are the most important types of tunable lasers for investigating biomolecular processes. Characteristic properties of typical pulsed and continuous wave (cw) dye lasers are given in Table 2; for details the reader is referred to the literature e.g. [1,2,3]. Optical frequency doubling, nonlinear mixing or Raman scattering (Fig. 1)[28] allow the extension of the spectral ranges. Extremely short pulses (picosecond range) are obtained with active mode-coupling by internal modulators or with passively mode-locking by saturable absorbers [2,20]. New observation instruments are also very important, e.g. streak cameras which enable the time-resolved detection of very short pulses.

Firstly, biomolecules can be examined in vitro in solution where conditions should resemble those inside a living cell — (ph 7 etc.). Secondly, they can be irradiated in living cells. In order to irradiate special sites inside a single cell, a laser beam is connected to a microscope (microbeam systems) [15,16]. Furthermore, dyes can be bound to special biomolecules, in order to observe e.g. the "secondary fluorescence" of the bound dye molecules if the biomolecules absorb only in the UV or if they have an extremely low quantum yield of fluorescence. Thirdly, questions of photo-

Table 2. Characteristic properties of tunable dye lasers with different pumping sources (cw = continuous wave)

Pumping Source	Tuning Range [nm]	Average Power [W]	Power [W]	Pulse length [ns]
Flashlamp	400– 800	0.1–100	10^5	10^2–10^5
N_2-Laser	350–1000	0,1– 1	10^4–10^5	1–10
Excimer-Laser	320– 980	0.4	10^4–10^6	1–10
YAG-Laser				
$\lambda/2$ = 530 nm	350– 800	0.1– 1	10^4–10^6	5–30
$\lambda/3$ = 355 nm				
$\lambda/4$ = 265 nm				
cw-Argon- or cw-Krypton-Laser	400– 800	0.1– 10	max. 40	cw

medical interest can also be answered directly on patients, e.g. by determining phototherapeutic action spectra (Chapter 6).

The experimental methods for the investigation of biomolecules cover a wide field such as fluorescence- and absorption measurements (flash photolysis, transient spectra), Raman-, picosecond- and Doppler spectroscopy. Examples of experimental apparatusses are given in Chapter 4 [8,17]. The various methods deliver information about different molecular properties. For example, fluorescence [18], fluorescence lifetimes and absorption measurements [43], especially in the picosecond range [20], inform about excited states of biomolecules and biochemical reactions. Raman spectroscopy [21,22] (Raman, resonance Raman and coherent anti-Stokes Raman scattering (CARS) [17]) as well as Rayleigh scattering experiments [23,24] and fluorescence correlation spectroscopy [25] give knowledge mainly about structural changes, conformation and motion of biomolecules. Special techniques on the basis of fluorescence, absorption and scattering parameters are developed for cell diagnosis of medical concerns: flow cytometry and cell analysis and sorting [26,27].

4 Investigations of Biomolecules

In this chapter typical examples of experimental arrangements and results of laser spectroscopy of biomolecules are given. Reference will be made to experimental details and for special results [8,9,10,18,19,29,30].

4.1 Chlorophyll

Chlorophyll is the most important light absorbing pigment in the process of photosynthesis. It consists of a porphyrin system with magnesium in the center. In photosynthesis of higher plants light energy is absorbed and transduced to chemical energy by the production of high energy phosphates. The primary step includes absorption by aggregates of pigments (photosynthetic unit) and a transfer of the absorbed energy to the reaction centers of photosystem I and II. The photosynthetic unit consists of the reaction center surrounded by large antennae of pigment molecules such as chlorophyll and carotinoids and the associated proteins, in which they are located. These units are found in the thylakoid membranes in the plant's chloroplast. The antennae increase the effective absorption cross section funneling the absorbed light energy to the reaction center [82].

The light absorption and energy migration to the reaction center takes place in a pisosecond time scale and can be monitored by observing the temporal behaviour of the fluorescence or absorption from different molecular species after picosecond light excitation.

Seibert and Alfano [31] investigated the time dependence upon the fluorescence of isolated spinach chloroplasts at room temperature. They used a frequency doubled, modelocked Nd:glass laser and an optical Kerr gate with an instrument resolution time of 10 ps. The excitation took place by 4 ps pulses at 530 nm and the fluorescence was observed at 685 nm (Fig. 5). With this time resolution, one observes two maxima in the fluorescent emission kinetics with a delay between them. The two peaks could be associated with components of photosystem I (lifetime of 10 ps) and photosystem II (lifetime of 210 ps). The 90 ps delay between the two peaks is related to the energy transfer between accessory pigments. These interpretations were based on theoretical models [31]. The best conformity with the experimental results was delivered by a model which considers two independent absorbing species, one is fluorescing and the other one transfers energy to a third fluorescing species. This suggests that the 90 ps delay of the second peak is due to an absorbing species at 530 nm (perhaps a carotenoid) which passes energy on to the photosystem II chlorophyll.

The fluorescence lifetimes of both photosynthetic systems and chlorophyll solutions using picosecond excitation and streak camera technique were also reported by Kolman et al. [32] and Shapiro et al. [33].

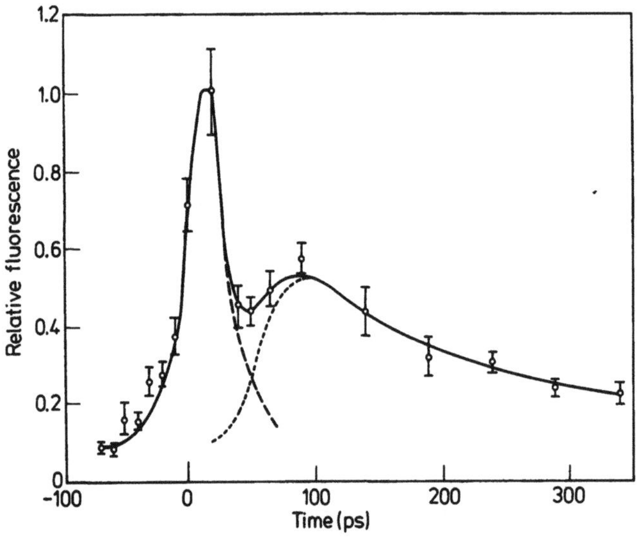

Fig. 5. Time dependence of fluorescent emission from spinach chloroplasts at 685 nm after excitation by a 4 ps pulse at 530 nm. Chlorophyll concentration: 35 µg/ml [31]

As all pigment molecules in the antennae can absorb the light energy there must be an efficient mechanism by which the energy is transferred without degradation to the reaction centre. This is described by a dipole-dipole interaction, the Förster energy transfer mechanism, which acts at a rate depending on the inverse sixth power of the distance between the donor and acceptor molecules, on their mutual orientation, and on the overlap between the absorption of the acceptor and the emission of the donor [e.g. [29,34]]. This excitation is suggested to be hopping randomly between neighbouring chlorophyll molecules ("random-walk-approach") and is usually referred to as exciton migration [29,35,82]. A general definition of a "biological exciton" is given by Knox [36]. The photosynthetic unit fulfills the condition of closeness and large spectral overlap for a rapid energy migration process. For this the detailed position of the pigments in the proteins may be important, but little is known about it at present. On the contrary, the manner in which excitons interact in vivo should inform about the topology of the photosynthetic unit.

Several experimental observations using single or multiple pulse laser excitation and streak camera detection [35,37] were concerned with the exciton migration in photosynthetic systems; a review is given by Campillo and Shapiro [35]. Campillo et al. [37] demonstrated, e.g. the singlet-singlet exciton annihilation occurring within the photosynthetic unit of the green alga Chlorella. The dominant mechanism is singlet-singlet fusion, manifesting itself by a decrease in the measured lifetime and quantum efficiency of fluorescence for high intensity single pulse excitation. Differences in low intensity (10^{13}–10^{16} photons/cm^2) and high intensity excitation (10^{17} photons/cm^2 and higher) and various quenching mechanisms are discussed [35].

Important information about the dynamics of bacterial reaction centres have been obtained from nanosecond and picosecond flash photolyses. Klevanik et al. [39] used a subpicosecond double-beam absorption spectrometer consisting of a passively mode locked cw dye laser pumped by an argon laser and an amplifier system. After amplification the beam was split into two parts, one of which was directed into a "white light" continuum by passing a cell containing H_2O and served as probing pulse, while the second one was used as an exciting pulse. The wavelength region for the exciting pulse was extended by stimulated Raman scattering in cyclohexane. The pigments were excited at 718 nm; with the aid of delay lines the changes in the absorption spectra after excitation were followed. The transfer of an electron from bacteriochlorophyll P_{870} to the primary acceptor could be measured with 7 ± 1 ps. Furthermore, the distance between different chromophores (bacteriopheophytin and bacteriochlorophyll P_{870}) could be estimated to be 12 ± 1 Å [39].

Optically-detected magnetic resonance spectroscopy (ODMR) has been used to study the chlorophyll triplet state. Clarke et al. [40] utilized an apparatus combining a pair of argon lasers with a microwave spectrometer. Preliminary results are dedicated to determinate structural dynamical features of chlorophyll in vivo. In such experiments the chlorophyll triplet state can be considered as a paramagnetic probe for the surroundings and interactions within the photosynthetic systems [40].

Resonance Raman spectra of spinach chloroplasts have been studied by Lutz and Breton [41]. The pigment constituents could be selectively excited by varying the exciting wavelength. Raman spectra from living cells of the alga Chlorella have been measured by Drissler [22,42] to get information about possible vibrational states in photosynthesis. A time-dependend Raman set-up with argon laser excitation

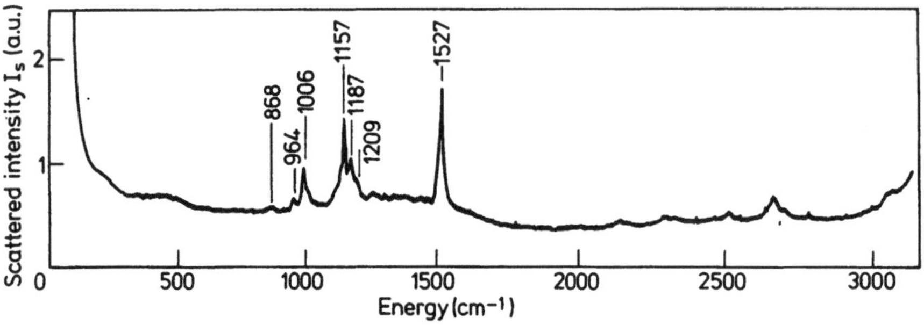

Fig. 6. Stokes Raman spectrum of Chlorella pyr. cells observed at room temperature during irradiation with 514.5 nm (19436 cm^{-1}) without additional background light (laser power: 1.0 W) [42]

was used. An example of a Stokes Raman spectrum is given in Fig. 6. The main features are 4 groups of lines (width $\sim 15\ cm^{-1}$) between 800 and 1600 cm^{-1}. These lines are assigned to carotinoid vibrations. It would be of interest whether a correlation between a biological function and a special spectral shape can be identified [22].

Concerning the search for new alternative energy sources, the exact knowledge on the mechanism of the primary steps in photosynthesis may create novel effective converters of solar energy; since the overall quantum efficiency of the primary photosynthetic steps is very high (about 0.9) [38].

4.2 Nucleic Acids and Nucleic-Acid-Dye-Complexes

4.2.1 Nucleic Acids

Nucleic acids store and transmit genetic information and regulate metabolism in living cells. The genetic code is represented in the sequence of bases in the desoxyribonucleic acid (DNA) helix (Fig. 10).

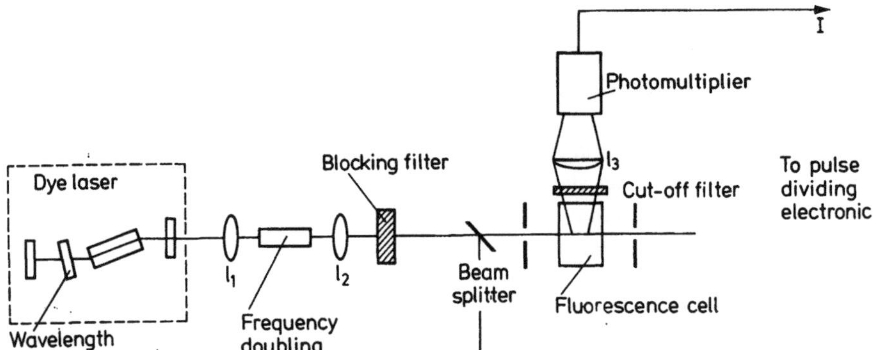

Fig. 7. Experimental arrangement to measure the fluorescence of biomolecules in solution after UV excitation with a pulsed dye laser [44]

The fluorescence of nucleic acids and of nucleic acid components (e.g. of the bases) is difficult to be measured at room temperature and at neutral pH, since the fluorescence quantum yield lies in the range 10^{-4} to 10^{-5}.

For this reason many investigations have been carried out at low temperatures. In order to examine the excited states of the nucleic acids with lasers, frequency doubled dye lasers or Nd:YAG lasers have to be used, because the nucleic acids absorb below 310 nm (Fig. 1, 2). An experimental arrangement for the observation of fluorescence in solution after excitation with a pulsed dye laser in the visible and UV down to 220 nm is shown in Fig. 7 [18,44].

The fluorescence spectrum of calf thymus DNA was measured by Anders [45] after excitation with a frequency doubled pulsed dye laser at 266 nm (Fig. 8); the quantum yield at room temperature was evaluated to about $2 \cdot 10^{-5}$.

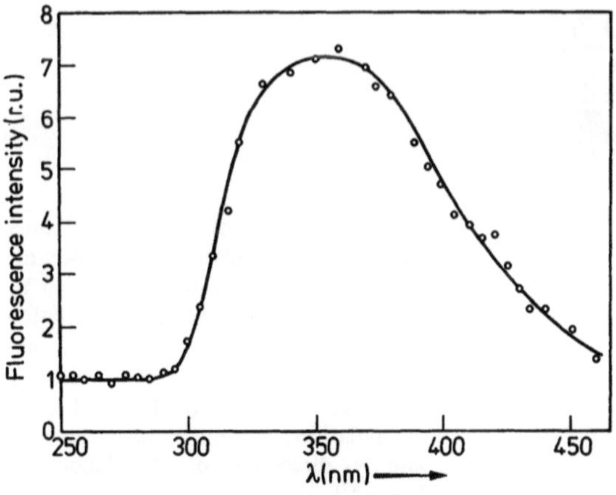

Fig. 8. Fluorescence spectrum of calf thymus DNA at room temperature in neutral aqueous solution. Exciting laser wavelength: λ = 266 nm [45]

Energy migration inside the nucleic acids, like an exciton transfer (see 4.1) along the DNA chain, and the deactivation of the electronic states proceed in the picosecond range [29]. Energy migration in nucleic acids plays an important role; e.g. for UV-produced defects in the DNA of cells and for the interaction between DNA and small molecules (see 4.2.2). Energy transfer along the DNA has been investigated in DNA-dye-complexes (see 4.2.2). Time resolved observations of the energy transfer in DNA-acridine-orange-complexes were performed by Shapiro et al. [46]. They excited the complexes in solution with 10 ps pulses from a mode-locked Nd:YAG laser at 265 nm. The measurement of the fluorescence risetime provides the interval over which the exciton transfer along the DNA takes place.

Experiments with frequency doubled pulsed dye lasers were done using dye-complexes with DNA of varying base contents, isolated from different organisms, and with synthetic polynucleotides of a fixed base sequence [44]. The energy transfer from DNA to the dye (see 4.2.2) in complexes, where the different mean spacings of the dye molecules along the DNA chain varied, rendered the range of the exciton transfer along the DNA. A dependence of the exciton transfer on the excitation wavelength and on the base composition of the DNA was found. The transfer increases near the 0—0 energy of DNA; this may be explained by a decrease of radiationless processes in big molecules when they are excited at their long-wavelength absorption edge.

The exciton transfer drops to 1/e, e.g. along 80 base pairs in Escherichia coli DNA (49.8 % GC) and along 60 base pairs in GC polynucleotides (100 % GC).

Applying the Förster theory (see 4.1) to the excitation transfer between the bases in a DNA chain a mean distance of the exciton transfer of about 55 base pairs was calculated.

Information about conformation and motion of nucleic acids was obtained by using pulsed fluorescence and fluorescence correlation spectroscopy [25,47]. A structural conformation can often be fingerprinted by the fluorescence lifetime of molecular fluorescing group or of an inserted label.

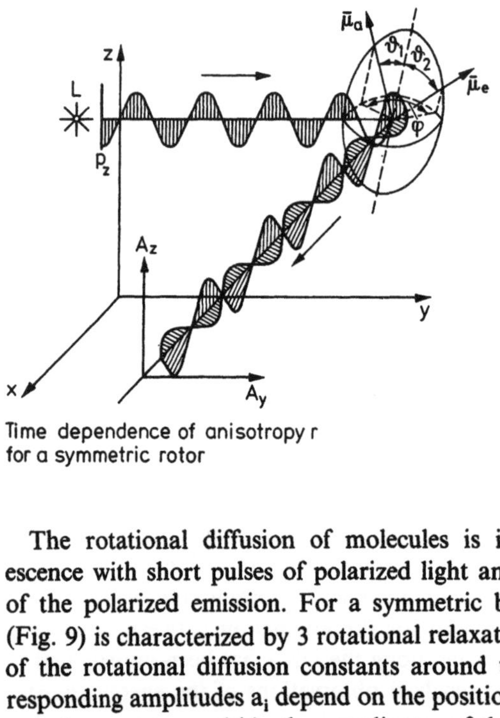

Fig. 9. Rotational diffusion of molecules as detected in pulsed fluorescence experiments. r(t): anisotrophy emitted by a symmetric rotor after pulsed excitation with polarized light L. μ_a and μ_e: absorption and emission vectors with coordinates 1, 2 and 0 (molecular frame). P_z: polarizer, A_z and A_y analyzers in z and y directions of the laboratory frame [47]

The rotational diffusion of molecules is investigated by exciting their fluorescence with short pulses of polarized light and by observing the time dependence of the polarized emission. For a symmetric body the anisotropy of the emission (Fig. 9) is characterized by 3 rotational relaxation times each of which is a function of the rotational diffusion constants around the main axes of rotation. The corresponding amplitudes a_i depend on the position of the absorption vector μ_a and the emission vector μ_e within the coordinates of the rotating unit.

The rotational relaxation time informs on problems like the sequental flexibility of antibodies, of DNA, of myosin, of membrane proteins or the rotational diffusion of transfer-RNA [47]. For example transfer-RNA (ribonucleic acid) which was labelled with the dye ethidium bromide was investigated using nanosecond laser pulses [25,60].

Transfer-RNA molecules participate in protein synthesis according to the genetic code at the ribosomes in the cell. The results indicate that conformational changes of the transfer-RNA molecules are important for the interaction between codon and anticodon at the ribosomes, for example. Two distinct and invariant lifetimes of the ethidium label in the anticodon loop of the transfer-RNA for the aminoacid phenylalanine in solution indicated two conformations [25].

4.2.2 DNA-dye-Complexes

The interaction of biologically active molecules like dyes with nucleic acids is of great interest because such molecules are used as drugs in chemotherapy (e.g. antibiotics, antitumor substances); other ones can induce mutations and tumors. They form intermolecular complexes with nucleic acids. A special class are dyes interacting with DNA by intercalation. These dye molecules are inserted between two base pairs (Fig. 10). A peculiarity of such complexes are base-sequence-dependent effects, for example, the fluorescence behaviour or the energy transfer inside the complex. These processes can depend on the type of base pairs between which the dye molecules are inserted. The quantum yield of fluorescence for some dyes after intercalation is much higher in AT-AT than in AT-GC or GC-GC sequences.

A base-sequence-dependent fluorescence lifetime was measured with a nitrogen laser pumped dye laser in quinacrine mustard-DNA complexes by Andreoni et al. [48,49]. The excitation took place at 419 nm corresponding to the absorption peak of quinacrine mustard. A nonexponential decay for AT complexes and an exponential decay for GC complexes (τ = 18 ns) was found (Fig. 11).

Base-sequence-dependent energy transfer processes from DNA-bases to intercalated dye molecules in solution were measured by Anders [44,51] using a frequency doubled pulsed dye laser (Fig. 7). Evidence for energy transfer from DNA-bases to intercalated dye molecules was already found by Weill and Calvin [52]. The quantum yield of the dye fluorescence after excitation in the visible and the UV range (where not only the dye but also the DNA absorb) render the relative amount of energy transfer from the bases to the intercalacted dye molecules. The energy transfer from the bases to the dyes (e.g. proflavine, acridine orange, ethidium bromid) depend on the excitation wavelength and on the kind of bases between which the dye molecules are inserted [44,51].

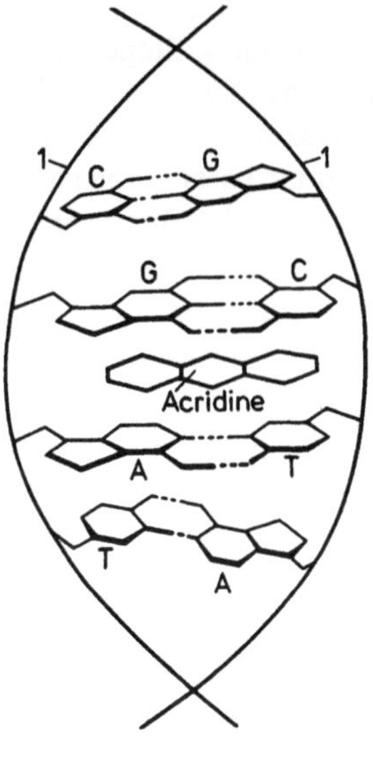

Fig. 10. Intercalation: Insertion of a planar dye molecule between adjacent base pairs in the DNA double helix. Three combinations are possible: insertion between two adenine thymine (AT) or guanine-cytosine (GC) pairs or an AT and a GC base pair sequence; 1: sugar-phosphat chain

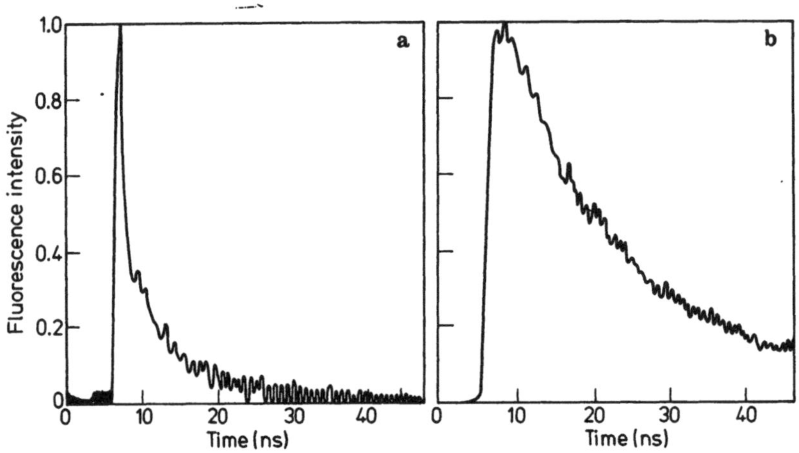

Fig. 11. Fluorescence decay of quinacrine bound to poly dG-poly dC (left) and poly dA-poly dT in solution (right) [50]

Such energy-transfer measurements also give information about molecular structures of the complexes, e.g. the position of the dye molecules in the plane between the bases when the transition dipole moments are considered [51]. The transfer from DNA to dye proceeds via a singlet-singlet mechanism. Since a singlet-singlet transfer predominates at room temperature, the radiationless deactivation by intersystem crossing is very low. In acridine-orange-DNA-complexes a maximum transfer for an excitation at 240 nm and around 300 nm, near the 0—0 energy of DNA at the long wavelength absorption edge, was observed. These results can be explained with energy transfer models considering all radiationless processes competing with the energy transfer and comparing the orientation of the transition dipole moments of the bases to those of the dye molecules in the different excitation regions [51].

A special class of DNA-dye complexes are the psoralen complexes (Chapter 6). They are used as drugs in the photochemotherapy of dermatosis such as psoriasis and vitiligo. Their biological action is attributed to their triplet states. Sloper et al. determined the quantum yield of the triplet formation by laser flash spectroscopy [53].

Another property of such dyes is observed when they are used to stain metaphase chromosomes. When these dyes bind to chromosomal DNA, some parts of the chromosome fluoresce stronger than others giving a typical and reproducible fluorescence pattern. These fluorescence are used for chromosome analysis and identification. Andreoni et al. [48–50] investigated the fluorescence decay of "in band" and "out of band" parts in Vicia fabia chromosomes stained with quinacrine mustards. They used a microbeam system (Chapter 3) with a nitrogen laser pumped dye laser. Their results seem to suggest that the band considered was to be attributed mainly to a higher dye concentration inside than outside the band.

4.2.3 General Remarks

Further experiments about DNA are discussed in Chapter 5.

As pointed out in Chapter 2 UV-light may cause lesions (like pyrimidine dimers) in DNA. An error made in the cell during the repair of damaged DNA could impact the first step towards *carcinogenesis* [54]. Therefore, the study of the molecular basis of UV radiation *mutagenesis* will remain an important area of interest in photobiology. As shown above the energy transfer along the DNA helix increases in that wavelength region (near 300 nm) where mutations occur from sunlight (Fig. 2).

Considering *chromosome staining* (Chapter 4.2.2), additional fluorescence bands or a disappearance of bands have been observed in connection with special diseases, e.g. with the cancer disease Burkitt lymphoma (probably caused by a virus). To elucidate such phenomena a selected laser excitation of special chromosome parts seems promising.

Furthermore, the investigation of the different action of intercalating dyes as they relate to carcinogenic versus *anticancerous properties* will be very important.

4.3 Hemoglobin, Myoglobin

The kinetics of fast reactions have been measured in hemoglobin, the oxygen carrier protein in red blood cells and in myoglobin, a muscle protein. In both proteins molecular oxygen is reversibly bound to the iron atom in the heme group of porphyrin structure. Furthermore, the cooperative binding of CO and other ligands to hemoglobin imposes a fundamental problem that has not been quantitatively solved. The mechanism of their reactivity seems to contain a sequence of structural and electronic events. Photodissociation experiments can inform about such changes during the reaction time.

Pico- and subpicosecond photodissociation was obtained by using mode-locked cw dye laser or a frequency doubled mode-locked Nd:glass laser [55–58].

A spectrometric set-up to measure transient spectra in the subpicosecond range is shown in Fig. 12. Half of the amplified pulses from the subpicosecond laser (L) is split into a cell of H_2O. The residual pulses pass through a variable delay line and are focused to excite the sample (beam 2). The delay is automatically scanned by a stepping-motor-driven stage (M). The emerging continuum beam is split into two parts (beams 0 and 1). Both are focused into the sample (E) and after that on the entrance slit of a spectrograph (S). The resulting two dispersed spectra are recorded by an optical multichannel analyser (D). The description of the passively mode-

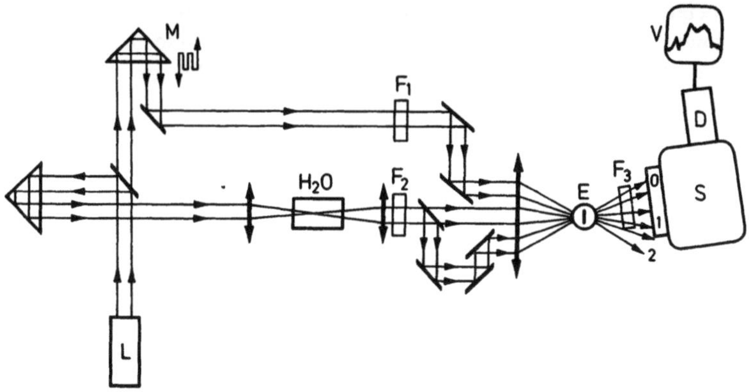

Fig. 12. Experimental set-up for measuring transient spectra in the subpicosecond range [58]

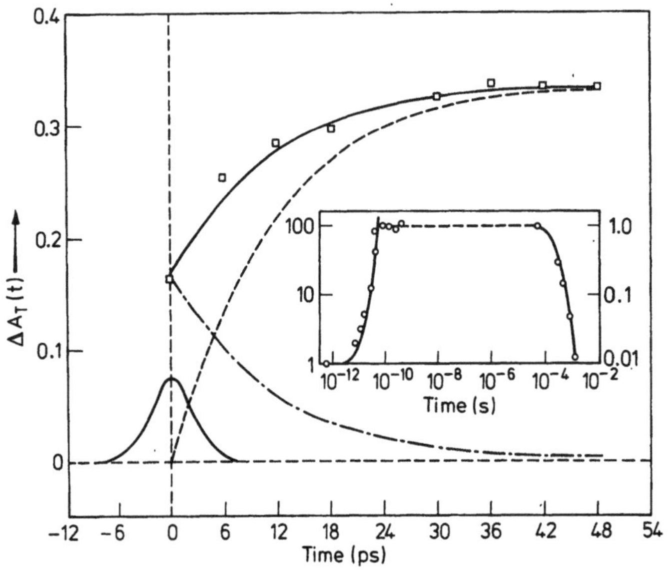

Fig. 13. Absorbance changes in HbCO at 440 nm from T = 0 to t = 48 ps after single-pulse (about 0.8 mJ) excitation at 530 nm in solution; □ experimental $A_T(t)$; ——— calculated $\Delta A_T(t)$; – – – calculated $\Delta A_C(t)$; —·— calculated $\Delta A_B(t)$. The excitation pulse profile is indicated at the origin. Inset: A log-log plot covering 10 orders of magnitude in time, shows the exponential kinetics of both the dissociation and the recombination initiated by a single 530-nm pulse. The left ordinate, A_{max}/I, $A_{max} - A_T(t)I$, corresponds to the photodissociation and the right ordinate, $A(t)/A_{max}$, corresponds to the recombination, — calculated line; ○ experimental points [57]

locked cw dye laser (0.7 ps) and a subpicosecond amplifier system is given by Martin et al. [58].

The absorbance change with carbon monoxide hemoglobin (HbCO) is demonstrated in Fig. 13 as a result obtained by Noe et al. [57]. In this experiment a modelocked Nd:glass laser at 530 nm (6 ps) was applied as excitation source. The data were displayed in the form ΔA, change in absorbance between excitation and no excitation, versus time (Fig. 13).

The photodissociation at 4 °C by single pulse excitation at 530 nm takes place in 11 ps and follows first-order kinetics. The kinetics of photodissociation, monitored by following absorbance changes in the Soret band at 440 nm were interpreted as corresponding to predissociation followed by a crossing into a dissociative state. Furthermore, subsequent recombination of CO with the porphyrin system and protein structural transformations were measured by use of a cw Helium-Cadmium-laser. These events occur in three distinct time regions depending on excitation pulse energy and pulse repetition rate [57].

Information about changes between different quaternary structures in the protein matrix of hemoglobin were obtained by Friedman et al. [59] and Lyons et al. [62]. The authors recorded resonance Raman spectra after photolysis of HbCO. They used a 10 ns Nd:YAG laser pulse at 530 nm for photolysis and monitored the Raman spectra of the transient species (occurring within 10 ns after photolysis) with a N_2-pumped dye laser.

Photodissociation was induced in carboxymyoglobin and oxymyoglobin with ps pulses from a Nd:YAG laser at 530 nm by Eisert et al. [61]. The photodissociation occurred in less than 8 ps; the kinetic and structural alterations were measured by following absorbance changes. The experiments showed that the mechanisms of photodissociation for carboxymyoglobin and oxymyoglobin are probably different. The relaxational differences between both kinds of molecules observed could be due to differences in tertiary structural changes of the heme pocket.

4.4 Rhodopsin

In the process of vision, light is absorbed by photoreceptor molecules like rhodopsin followed by a formation of several intermediate products (Fig. 14). Rhodopsin consists of an 11-cis retinal chromophore and opsin, a protein. After the absorption of a photon the 11-cis bond of retinal is transformed into a 11-trans bond.

The first photochemical events in vision under physiological conditions have to be studied with picosecond spectroscopy because of the rapid formation [62,63,64]. Earlier observations have been made at low temperature. The formation lifetime of the first intermediate product (prelumirhodopsin) was measured to be less than 6 ps at room temperature [63]. A double-beam picosecond spectrometer with a mode-locked Nd:YAG laser was used [62].

Rentzepis [64] suggests a proton tunneling effect (e.g. in the Schiff base retinal) for the mechanism of prelumirhodopsin formation, at least at low temperatures. The ultrafast rate observed for the formation of prelumirhodopsin could be interpreted with an activation without a barrier for the proton translocation. The distance, through which the proton tunnels could be calculated to be 0.5 Å [62]. Models for the proton translocation are also discussed by Applebury [82].

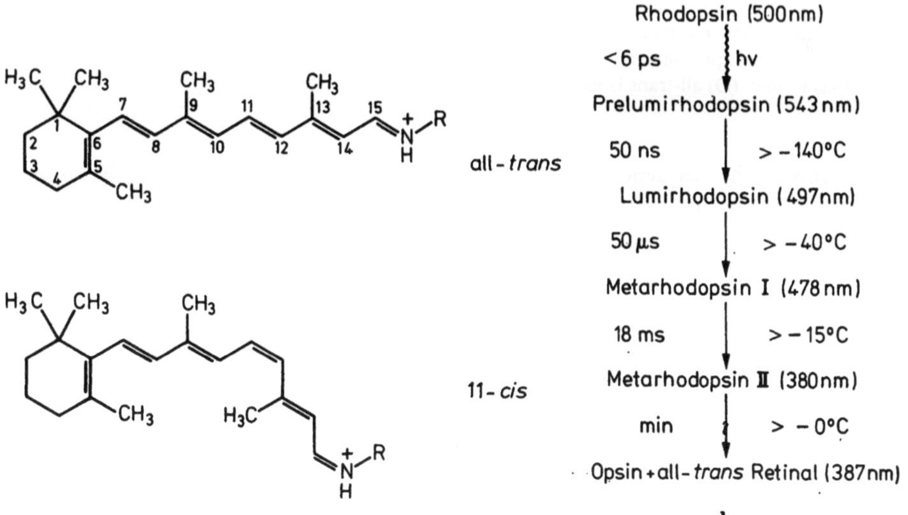

a b

Fig. 14. **a** Protonated Schiff bases of all-trans and 11-cis retinal. **b** Sequence of intermediates in the photolysis of rhodopsin. The absorption maximum of each intermediate is shown in parenthesis [65]

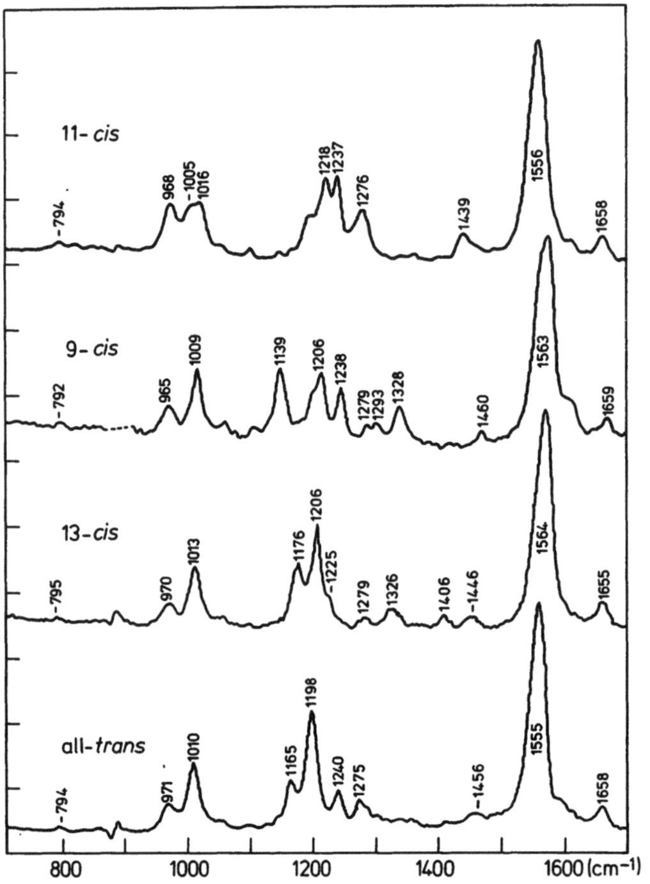

Fig. 15. Rapid-flow resonance Raman spectra of protonated Schiff bases of retinals in ethanol (chloride salts of protonated n-butylamine derivatives). (A) 11-cis isomer. (B) 9-cis isomer. (C) 13-cis isomer. (D) all-trans isomer [65)]

To reveal the sequence of conformational changes occurring in the retinal chromophore group and in the protein, Raman spectra were recorded [65,66,67)]. Resonance Raman spectra of the intermediates in the photolysis of rhodopsin are given in Fig. 15. The spectra were measured with cw dye lasers using a rapid-flow technique to obtain the spectra before a photolabile molecule is altered by light [65)]. The differences in the fingerprint region (1100–1350 cm^{-1}) demonstrate the sensitivity of the vibrational spectrum to the conformation.

4.5 Miscellaneous Topics

Laser *light scattering* applied to biomolecules is described by Ware, Giglio and Morris et al. [24,69,23)]. A common experiment of biological interest is the analysis of the angular dependence of the intensity of scattered laser radiation from *macromolecules* or molecular aggregates in solution.

Quasi-elastic laser light scattering (also called intensity fluctuation spectroscopy, light-beating spectroscopy or photon correlation spectroscopy) is an accurate method to measure the translational *diffusion coefficients* of macromolecules. The diffusion coefficient is a parameter, that depends on the size and shape of the macromolecules and on the thermodynamic and hydrodynamic interaction between the macromolecules.

Radiation which is scattered from a moving particle is shifted in frequency by the Doppler effect; the magnitude of the Doppler shift allows an exact determination of the velocity of the object. *Laser velocimetry* in biology and medicine is reviewed by Ware [24].

A further example are Laser Raman spectroscopy resonance studies of the enzyme aldolase catalyzing a key reaction in the muscle cells. Zerbi et al. [71] investigated resonance Raman spectra of labelled aldolase with argon ion and krypton ion lasers.

The complex formation of bilirubin with human serum albumin was investigated by Sinclair et al. [72] using 347 nm ruby laser flash photolysis technique. A high bilirubin level is found in new born babies who suffer from jaundice (neonatal *hyperbilirubinemia*) [73]. *Phototherapy* has been found to be suitable for lowering the bilirubin level. In order to understand the mechanism of the phototherapy, investigation into the photophysics of bilirubin is essential. It is strongly bound to human serum albumin, lipids and cell membranes.

A new technique for measuring equilibrium adsorption/desorption kinetics and surface diffusion of fluorescent-labelled solute molecules at surfaces was developed by Thompson et al. [74]. The technique combines total internal reflection fluorescence with either fluorescence photobleaching recovery or fluorescence correlation spectroscopy with lasers. For example, fluorescent labelled protein was studied in regard to the surface chemistry of blood [75].

Information on transport processes through membranes of red blood cells were obtained by fluorescence investigations using a nitrogen pumped dye laser as excitation source by Marowsky et al. [76]. The membrane potential-dependent uptake and release of dye, which was added to the solution of the blood cells, was observed.

The action spectrum of photoreactivation (in yeast cells) was measured by Anders et al. [77], using different pulsed dye lasers. Photoreactivation occurs as a repair process in DNA after defects caused by short UV wavelengths. The repair processes are not completely understood [54] (Chapter 4.2.3).

5 Selective Excitation

Tunable lasers with their high spectral intensity selectively excite single quantum states in atoms and molecules in a wide range in the visible, ultraviolet and infrared part of the spectrum. Selective laser action was successfully applied in photophysics and photochemistry [78,79]. Their applications on biomolecules could, in the future, achieve great importance in the control of complicated biochemical reactions. Selective laser stimulation acts mainly in two ways:

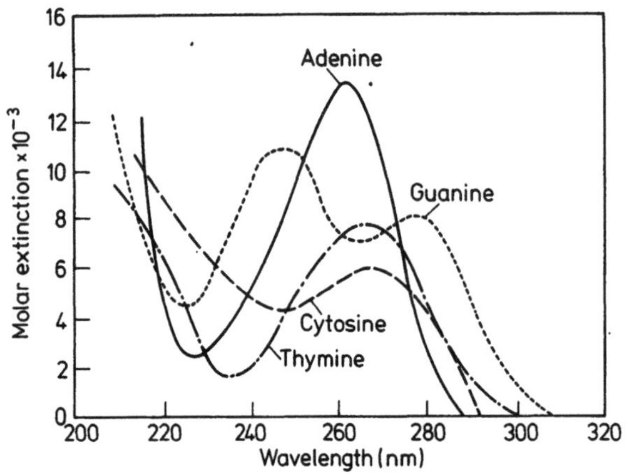

Fig. 16. Absorption spectra of DNA-bases [14)]

a) The excitation of special molecules in a mixture of different molecules similar to isotope separation.
b) The excitation of single energy levels in one kind of atoms or molecules, in order to control the photochemical reaction of such a selectively excited molecule.

The selective excitation in complex biological molecules does however, present considerable difficulties because of the wide and overlapping absorption bands (e.g. Fig. 16; DNA bases) and because of the very fast relaxation and intramolecular energy transfer. Because of a little difference in the absorption spectra, differences in absorption cross-sections from the excited states, lifetimes of the excited states or differences in rates of photoinduced chemical reactions can be used. Apart from a one-step electronic excitation two- or more-step excitations can be carried out (Fig. 17). In case of a IR-UV-excitation the first step must be done in picoseconds in order to obtain the selectivety for the second step, because of the very fast vibrational relaxation [78,79)]. Considering future experiments in living cells the problem of IR absorption by the water in the cells has to be solved.

Some experiments using selective laser excitation on biomolecules have been started. In Table 3 selective processes in nucleic acids are summarized.

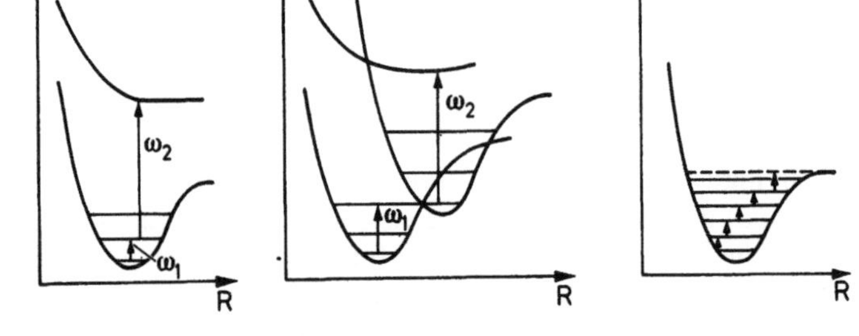

Fig. 17. Typical molecular potential functions with types of selective photoexcitation: from left: 1) 2-step IR-UV excitation; 2) 2-step excitation via an electronic state; 3) multiphoton IR excitation

Table 3. Excitation of selective processes in nucleic acids with lasers

Molecule	Laser	Excitation	Observation	Ref.
Bases of Nucleic Acids	Nd: YAG Laser 266 nm 30 ps	Mixture of Bases 2-step	Photoproducts	Angelov et al. [80,81]
Poly dG-dC and Poly dA-dT Proflavine-complexes	N_2-pumped dye laser 430 nm; N_2-laser 337 nm	Photo-ionization Proflavine 2-step	Proflavine-damage Fluorescence-decrease	Adreoni et al. [82,83]
DNA- and Poly-acridine orange-complexes	Frequency-doubled dye laser 260–310 nm	Kinds of Bases in DNA 1-step	Interaction of DNA and Acridine orange, Energy Transfer	Anders [84]

Selective action on nucleic acid components in solution by picosecond pulses and the production of irreversible photoproducts were reported by Angelov et al. [80,81]. They used the fourth harmonic wavelength of a mode-locked Nd:YAG laser at 266 nm near the maximum of the first electronic absorption band of nucleic acids; the irradiation intensity amounted to about 1 GW/cm^2. Figure 18 shows the dependence upon the photoproduct yield versus irradiation intensity. The photoproduct yield was determined by the relative change in optical density. The molecular character of action was found to be different for each type of bases [81].

Andreoni et al. [49,83] investigated a selective photodamage of dye molecules bound to polynucleotides. The experiments were based on a time-delayed two-step photoionization; the experimental set up is shown in Fig. 19. The two-step excitation was performed with a nitrogen laser (337 nm) and a dye laser (pumped by the nitrogen laser) at 430 nm. The selectivity arises from the fact that the lifetime of the first excited electronic state of the dye may be sensitive to the binding site (Chapter 4.2.2). Two complexes with different lifetimes τ_1 and τ_2 ($\tau_1 < \tau_2$) are irradiated with two laser pulses (Fig. 20). The second pulse is delayed with τ_D against the first pulse. If $\tau_1 < \tau_D < \tau_2$ the second pulse (frequency ν_2) will find a smaller singlet-state

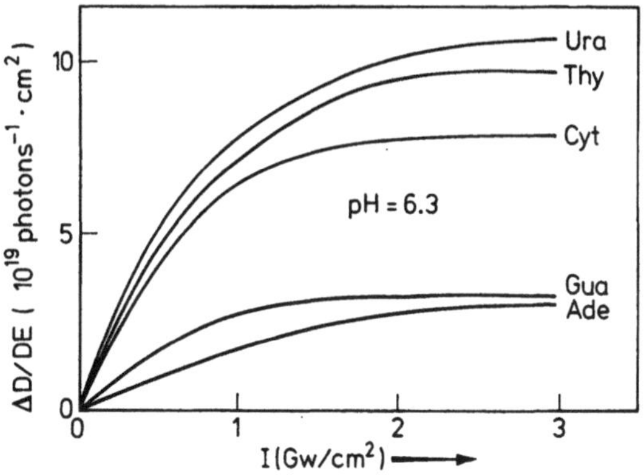

Fig. 18. Dependences of photoproduct yield versus irradiation intensity for all five DNA bases in neutral aqueous solution [81] (see text)

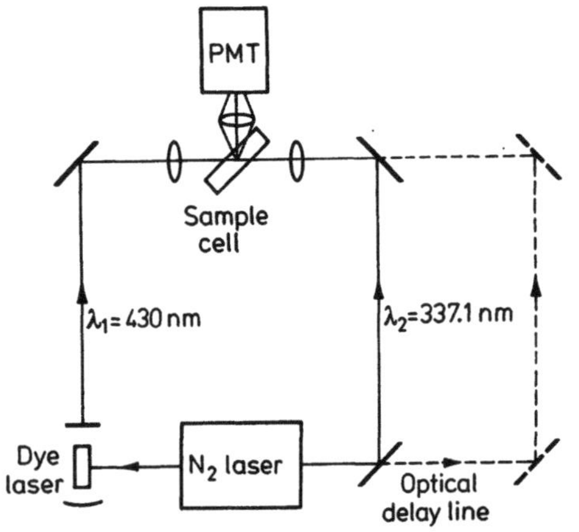

Fig. 19. Block diagram for a two-step photo-ionization [83]

population of complex 1 due to its faster decay. Thus, complex 2 will be more damaged. If $\tau_D < \tau_T$ (τ_T: triplet lifetime), complex 1 will be more damaged. Since $\tau_1 < \tau_2$, the intersystem crossing rate and therefore the triplet-state population will be larger for complex 1 than for complex 2. Assumably the ground-state absorption cross section and the triplet-state lifetime τ_T are the same for the two complexes [83].

The authors found a different damage probability for the dye proflavine bound either to poly AT or to poly GC. The damage was measured by the reduction of the fluorescence when the complexes were excited at 430 nm alone.

The favoured excitation of DNA bases (e.g. only, A, T, G or C) via a one-step electronic excitation was performed by Anders [84] and was tested by the interaction of these bases with intercalated dye molecules. In such DNA-dye-complexes energy is transfered from the bases to the dye and from base to base depending on the base sequence (see Chapter 4.2.2). In order to analyze selective effects the energy transfer in the three possible base pair units (Fig. 10) after the favoured excitation of one kind of bases was compared in different complexes. Two tendencies are found:

1) The energy transfer in a kind of base pair units is larger if mainly these units are excited. E.g., with ethidium bromide intercalated between AT—AT units (Chapter 4.2.2) the transfer is stronger in that region in which more AT than GC pairs are excited.

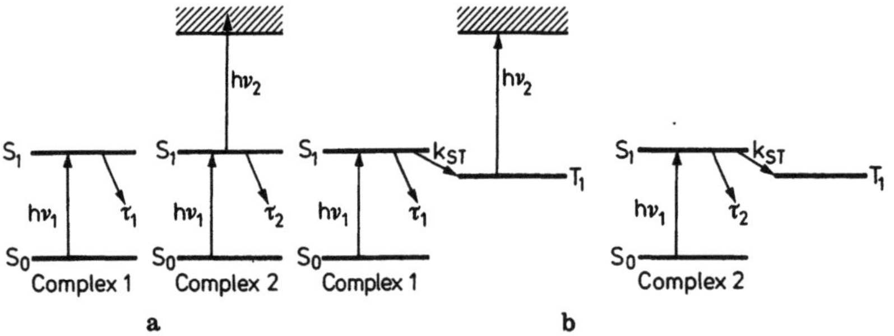

Fig. 20. Electronic states of complexes 1 and 2; a) selective ionization of complex 2; b) selective ionization of complex 1 [83]

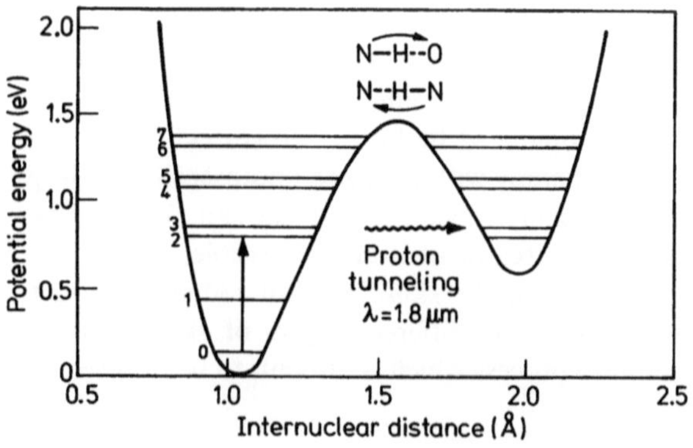

Fig. 21. Potential function of hydrogen bonds in DNA and possible method for excitation of levels by IR laser radiation to stimulate proton tunneling [79]

2) A higher transfer appears in some units if other than the considered ones are selectively excited. E.g., proflavine inserted in AT—AT units shows a greater transfer if GC pairs are excited and the energy is transferred along the DNA to the AT pairs.

The control of the replication process in DNA or an intended induction of mutation using laser excitation could be of great importance in the future. The selective breaking of hydrogen bonds in DNA resulting in a splitting of the helix is discussed by Letokhov [79]. Two pairs of bases in DNA have slightly different hydrogen bonds (Fig. 10). The G—C bonds correspond to an infrared absorption band at about 1720 cm^{-1} and the A—T bonds to a band of about 1700 cm^{-1}. Figure 21 gives the potential function of hydrogen bonds in DNA with calculated energy levels of a proton in the bond. Letokhov proposed the stimulation of hydrogen bond breaking with powerful picosecond IR pulses.

6 Photomedicine

The use of lasers in medical practice has gained growing importance [8,85]. Many of these applications rely on the high intensity and small focus as in surgery and in coagulation. A new class of laser applications in medicine represents the use of low-intensity-radiation [9,87]. In non-thermal use, the laser properties playing major part are high spectral intensity and tunability of the wavelength.

The use of lasers for diagnostic purposes is well established. It includes various techniques like cytofluorometry [26,88] (e.g. cell sorting and counting), Doppler techniques (e.g. measurement of blood flow), various kinds of spectroscopy, and holography [86].

The Raman spectroscopy in vivo to monitor the respiratory gases is one example of applying laser spectroscopy in medicine. This technique can be routinely used e.g. in clinical application for anesthetic control during operations. Another example is the laser nephelometry; it allows the determination of different protein concentrations in the blood by measuring the Mie scattering [2,88]. With a laser fluorescence bronchoscope very early lung cancer processes could be localized. The tumor was marked with hematopophyrin, which accumulates in malignent tissue to a greater extent than in normal tissue. The stained tumors were irradiated by a krypton ion laser using fiber optic technique and a bronchoscope. The fluorescence light of the tumor tissue was imaged in comparison to the surrounding dimmer tissue [89].

An improved knowledge about photoinduced biological processes in man is of great interest as a basis for the medical therapy with light and photosensitizing drugs. Photomedical therapy is, of course, especially concerned with the skin, but via fiber-optic techniques and endoscopes internal treatment is also accessible.

Figure 22 shows the optical properties of the skin. Incident radiation is partly reflected, partly absorbed or scattered in the different layers of skin. Thus, the action of light on skin is a very complex process which is difficult to be resolved into its various basic mechanisms [90,91].

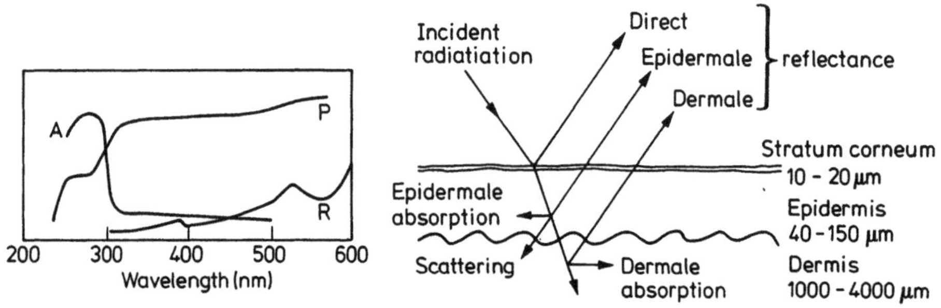

Fig. 22. Left: spectral distribution of absorption (A) and reflection (R) of radiation in skin, and depth of penetration (P) of light into the skin, measured with spectrophotometers. right: absorption, reflection, and scattering of optical radiation in skin (schematic) [87,90]

Spectroscopic properties of isolated human epidermis were measured with dye lasers. Structures in transmission or absorption spectra could be resolved, which were not found with usual spectrophotometers [92]. The optical behaviour of isolated skin treated with photosensitizing dyes was also investigated with tunable lasers; e.g. changes in transmission after irradiation with different laser wavelengths were detected [92,93]. Such dyes are used in photochemotherapy of dermatosis and tumors.

Photosensitizing dyes work through different mechanisms in the cell. For example, psoralens (Chapter 4.2.2) react under long-wavelength UV-irradiation with the thymine in DNA by cyclobutane addition forming mono- and diadducts. In this way the synthesis of nucleic acids is blocked. Pathological cell growth, as in the skin disease psoriasis, is thus prevented.

In many kinds of skin diseases a therapy with UV-light is applied; the same lamps are used for cosmetic tanning. Therefore, the different action spectra should be known with high accuracy to avoid unwanted effects. Such action spectra of medical interest are the erythema curve (the action spectrum of sunburn) and action spectra after oral or local administration of photosensitizers.

Frequency doubled dye lasers with their high spectral intensity and tunability are ideal light sources to get an improved knowledge about such UV-induced effects in skin. The erythema effectiveness curve was investigated with frequency doubled pulsed dye lasers by Anders et al. [87,94,95].

New understanding about the optical behaviour of the skin and DNA damage and repair can enter into the design of new therapies including photochemical reactions with known action spectra and metabolic consequences [54,90].

7 Conclusion

The application of laser spectroscopy in biology and medicine gained a remarkable importance and is still rapidly growing. Some future aspects of biomolecular laser spectroscopy are already mentioned in the respective chapters. New technologies which combine lasers with conventional instruments, an optical mass spectrometer and a laser ion microscope are discussed by Letokhov [79].

Extension of laser investigations to further biomolecular processes will proceed parallel with an improved technological laser development. In particular, reliable, push-button tunable lasers in the UV providing tunable UV-radiation with sufficient intensity, will greatly enhance the field of applications.

Spectroscopic investigations in the UV concern areas of photobiology with an impact on photomedicine, e.g. light activation of enzymes, UV-radiation effects on the immune system and on cellular membranes, the interaction of drugs with DNA (Chapter 4.2.2 and 4.2.3) and the role of DNA damage and repair in mutagenesis and carcinogenesis.

8 References

1. Walther, H., Rothe, K. W. (eds.): Laser Spectroscopy, Heidelberg: Springer 1979
2. Demtröder, W.: Laser Spectroscopy, Heidelberg: Springer 1981
3. Schaefer, F. P. (ed.): Dye Lasers, Heidelberg: Springer 1977
4. Kroon, D. J. (ed.): 1980 Europ. Conf. on Optical Systems and Applications SPIE *236*, 1981
5. Moore, C. B. (ed.): Chemical and Biochemical Applications of Lasers, Vol. 1, 2, 3, 4, Heidelberg: Springer (1974, 1977, 1980)
6. Kompa, K. L., Smith, S. D. (eds.): Laser-induced Processes in Molecules, Heidelberg: Springer 1979
7. Eichler, J., Lenz, H.: Appl. Opt. *16*, 27 (1977)
8. Hillenkamp, F., Pratesi, R., Sacchi, C. A. (eds.): Lasers in Biology and Medicine, New York: Plenum 1980
9. Pratesi, R., Sacchi, C. A. (eds.): Lasers in Photomedicine and Photobiology, Heidelberg: Springer 1980
10. Anders, A.: in ref. [4], pp. 160
11. Smith, K. C. (ed.): The Science of Photobiology, New York: Plenum 1977
12. Giese, A. C.: in ref. [9], pp. 26
13. Wells, C. H. J.: Introduction to Molecular Photochemistry, London: Chapman and Hall 1972
14. Hillenkamp, F.: in Ref. [8], pp. 37
15. Berns, M. W.: in Laser Applications in Medicine and Biology, Vol. 2 (ed. Wolbarsht, M. L.), New York: Plenum 1974, pp. 1
16. Berns, M. W.: Methods in Cell Biology (eds. Stein, G., Stein, J., Kleinsmith, L.), New York: Academic Press 1978
17. Hudson, B. S.: Ann. Rev. Biophys. Bioeng. *6*, 135 (1977)
18. Anders, A.: in ref. [9], pp. 158
19. McCray, J. A., Smith, P. D.: in Laser Application (ed. Ross, M.), New York: Academic Press 1977, pp. 1

20. Shapiro, S. L. (ed.): Ultrashort Light Pulses, Heidelberg: Springer 1977
21. Spiro, T. G.: in ref. [5], Vol. 1, pp. 29
22. Drissler, F.: in ref. [85], pp. 113
23. Morris, S. J., Shultens, H. A., Hellweg, M. A., Striker, G., Jovin, T. M.: Appl. Opt. *18*, 303 (1979)
24. Ware, B. R.: in ref. [5], Vol. 2, pp. 199
25. Ehrenberg, M., Rigler, R.: in Tunable Lasers and Applications (ed. Mooradian, A., Jaeger, T., Stokseth, P.), Heidelberg: Springer 1976
26. Melamed, M. R., Mullaney, P. F., Mendelsohn, M. L. (eds.): Flow Cytometry and Sorting, New York: Wiley 1979
27. Horan, P. K., Wheeles, L. L. jr.: Science *198*, 149 (1977)
28. Wilke, V., Schmidt, W.: Appl. Phys. *18*, 177 (1979)
29. Campillo, A. J., Shapiro, S. L.: in ref. [20], pp. 364
30. Parson, W. W.: in ref [5], Vol. 1, pp. 339
31. Seibert, M., Alfano, R. R.: Biophys. J. *14*, 269 (1974)
32. Kolman, V. H., Shapiro, S. L., Campillo, A. J.: Biochem. Biophys. Res. Commun. *63*, 917 (1975)
33. Shapiro, S. L., Kolman, V. H., Campillo, A. J.: FEBS Lett. *54*, 358 (1975)
34. Beddard, G. S.: in ref. [8], pp. 225
35. Campillo, A. J., Shapiro, S. L.: Photochem. Photobiol. *28*, 975 (1978)
36. Knox, R. S.: in Bioenergetics of Photosynthesis, (ed. Govindjee), New York: Academic Press 1975, pp. 183
37. Campillo, A. J., Shapiro, S. L., Kolman, V. H., Winn, K. R., Hyer, R. C.: Biophys. J. *16*, 93 (1976)
38. Rubin, L. B.: in ref. [9], pp. 221
39. Klevanik, A. V., Kryukov, P. G., Matveets, Yu. A., Semchishen, V. A., Shuvalov, V. A.: JETP Lett. *32*, 97 (1980)
40. Clarke, R. H., Jagannathan, S. P., Leenstra, W. R.: in ref. [9], pp. 171
41. Lutz, M., Breton, J.: Biochem. Biophys. Res. Comm. *53*, 413 (1973)
42. Drissler, F., Macfarlane, R. M.: in ref. [9], pp. 189
43. Truscott, T. G.: Photochem. Photobiol. *35*, 867 (1982)
44. Anders, A.: Opt. Commun. *26*, 339 (1978)
45. Anders, A.: Chem. Phys. Lett. *81*, 270 (1981)
46. Shapiro, S. L., Campillo, A. J., Kolman, V. H., Goad, W. B.: Opt. Commun. *15*, 308 (1975)
47. Rigler, R., Grasselli, P.: in ref. [8], pp. 151
48. Andreoni, A., Sacchi, C. A., Svelto, O.: in ref. [5], Vol. 4
49. Andreoni, A., Cubeddu, R., De Silvestri, S., Svelto, O., Bottiroli, G.: in ref. [9], pp. 216
50. Andreoni, A.: in ref. [8], pp. 165
51. Anders, A.: Appl. Phys. *18*, 333 (1979)
52. Weill, G., Calvin, M.: Biopolymers *1*, 401 (1963)
53. Sloper, R. W., Tuscott, I. G., Land, E. J.: Photochem. Photobiol. *29*, 1025 (1979)
54. Smith, K. C.: J. Invest. Dermat. *77*, 2 (1981)
55. Shank, C. V., Ippen, E. P., Bersohn, R.: Science *193*, 50 (1976)
56. Greene, B. I., Hochstrasser, R. M., Weisman, R. B., Eaton, W. A.: Proc. Nat. Acad. Sci. USA *75*, 5255 (1978)
57. Noe, L. J., Eisert, W. G., Rentzepis, P. M.: ibid. *75*, 573 (1978)
58. Martin, J. L., Astier, R., Migus, A., Antonetti, A.: in ref. [9], pp. 223
59. Friedman, J. M., Lyons, K. B.: in ref. [9], pp. 195
60. Ehrenberg, M., Rigler, R., Wintermeyer, W.: Biochem. *18*, 4588 (1979)
61. Eisert, W. G., Degenkolb, E. O., Noe, L. J., Rentzepis, P. M.: Biophys. J. *25*, 455 (1979)
62. Lyons, K. B., Friedman, J. M., Fleury, P. A.: Nature *275*, 565 (1978)
63. Busch, G. E., Applebury, M. L., Lamola, A. A., Rentzepis, P. M.: Proc. Nat. Acad. Sci. USA *69*, 2802 (1972)
64. Rentzepis, P. M.: Biophys. J. *24*, 272 (1978)
65. Mathies, R., Oseroff, A. R., Freedman, T. B., Stryer, L.: in Tunable Lasers and Applications (ed. Mooradian, A., Jaeger, T., Stokseth, P.), Heidelberg: Springer 1976
66. Marcus, M. A., Lewis, A.: Photochem. Photobiol. *29*, 699 (1979)

67. Aton, B., Doukas, A. G., Narva, D., Callender, R. H., Dinur, U., Honig, B.: Biophys. J. *29*, 79 (1980)
68. Ippen, E. P., Shank, C. V., Lewis, A., Marcus, M. A.: Science *200*, 1279 (1978)
69. Giglio, M.: in ref. [8], 111
70. Pearlstein, R. M.: Photochem. Photobiol. *35*, 835 (1982)
71. Zerbi, G., Masetti, G., Nannicini, L., Dellepiane, G.: in ref. [9], 182
72. Sinclair, R. S., Sloper, R. W., Truscott, T. G.: in ref. [9], 153
73. Vogl, T. P.: in ref. [9], pp. 136
74. Thompson, N. L., Burghardt, T. P., Axelrod, D.: Biophys. J. *33*, 435 (1981)
75. Burghardt, T. P., Axelrod, D.: Biophys. J. *33*, 455 (1981)
76. Marowsky, G., Cornelius, G., Rensing, L.: Opt. Commun. *22*, 361 (1977)
77. Anders, A., Yasui, A., Zacharias, H., Lamprecht, I., Laskowski, W.: Radiation and Cellular Control Processes, Heidelberg: Springer 1976, pp. 221
78. Letokhov, V. S.: in Laser Spectroscopy IV (ed. Walter, H., Rothe, K. W.), Heidelberg: Springer 1979, pp. 504
79. Letokhov, V. S.: Sov. Phys. Usp. *21*, 405 (1978)
80. Angelov, D. A., Kryukov, P. G., Letokhov, V. S., Nikogosyan, D. N., Oraevsky, A. A.: in ref. [9], pp. 207
81. Angelov, D. A., Kryukov, P. G., Letokhov, V. S., Nikogosyan, D. N., Oraevsky, A. A.: Appl. Phys. *21*, 391 (1980)
82. Applebury, M. L.: Photochem. Photobiol. *32*, 425 (1980)
83. Andreoni, A., Cubeddu, R., De Silvestri, S., Laporta, P., Svelto, O.: Phys. Rev. Lett. *45*, 431 (1980)
84. Anders, A.: Appl. Phys. *20*, 257 (1979)
85. Waidelich, W. (ed.): Laser 81 Opto-Electronics, Heidelberg: Springer 1982
86. v. Bally, G. (ed.): Holography in Medicine and Biology, Heidelberg: Springer 1979
87. Anders, A., Aufmuth, P.: in Laser 81 Opto-Electronics (ed. Waidelich, W.), Heidelberg: Springer 1982, pp. 136
88. Goldman, L.: in Laser 79 Opto-Electronics (ed. Waidelich, W.), Guildford: IPC Science and Technology, pp. 327
89. Doiron, D. R., Profio, A. E.: in ref. [9], pp. 92
90. Anderson, R. R., Parrish, B. S., Parrish, J. A.: J. Invest. Dermat. *77*, 13 (1981)
91. Magnus, I. A.: Dermatological Photobiology, Oxford: Blackwell Scientific Publ. 1976
92. Anders, A., Lamprecht, I., Schaefer, H., Zacharias, H.: Arch. Dermatol. Res. *255*, 211 (1976)
93. Anders, A., Zacharias, H., Aufmuth, P.: Laser 77 Opto Electronics (ed. Waidelich, W.), Guildford: IPC Science and Technology 1979, pp. 520
94. Anders, A., Aufmuth, P., Böttger, E.-M., Tronnier, H.: Dermatosen, in press (1984)
95. Anders, A., Aufmuth, P., Böttger, E.-M., Tronnier, H.: in ref. [9], pp. 83

Ion Pair Chromatography on Reversed-Phase Layers

Dieter-Günter Volkmann

Quality Control Laboratory, BAYER AG, Friedrich-Ebert-Straße 217–319,
D-5600 Wuppertal 1, FRG

Table of Contents

1 Introduction 52

2 Theory of Ion Pair Formation 52

3 The Chemistry of Reversed-Phase Layers 53

4 Mechanism of Ion Pair Chromatography on Reversed-Phase Layers 55

5 Techniques of Ion Pair Chromatography 56

6 Literature Survey 57

7 Ion Pair Chromatography on Impregnated Layers 57

8 Ion Pair Chromatography on Chemically Modified Layers 62
8.1 Chromatography on Silanized Silicagel 62
8.2 Chromatography on RP_2-, RP_8-, and RP_{18}-Layers 65

9 Summary and Outlook 68

10 References 68

1 Introduction

The analysis of ionogenic samples by thin layer chromatography is often difficult because an equilibrium exists between the ionized and non ionized forms in the mobile phase.

This particularly applies to the chromatography on reversed-phase layers as mobile phases consisting of hydrous-organic mixtures are mainly used.

Therefore spots on the chromatogram show tailing or are splitted into two spots. Two steps are possible to prevent this:

1. In the case of an acid one can add a stronger acid to the mobile phase (e.g. dilute hydrochloric acid or in some cases acetic acid). Where bases are concerned the addition of a stronger base (e.g. dilute ammonia or alkylamines) is advantageous. In both cases the equilibrium is shifted towards the unionized forms.
2. The addition of a counter-ion to the acid or base results in the formation of an ion pair. This counter-ion is provided with more lipophilic properties, thus influencing the chromatographic behaviour in a more defined manner.

It is the aim of the following chapters to give a short introduction into the theory and to illustrate the present situation of ion pair chromatography on reversed-phase layers. The great success achieved with reversed-phases in column chromatography has resulted in the use of reversed-phase sorbents also in TLC.

2 Theory of Ion Pair Formation

An ion pair is formed from a water soluble cation $[Q^{\oplus}]$ and a water soluble anion $[X^{\ominus}]$. The resulting ion pair is soluble in organic solvents but insoluble in water.

$$Q^{\oplus}_{aq} + X^{\ominus}_{aq} = QX_{org.} \tag{1}$$

The equilibrium of this reaction is controlled by the equilibrium constant (E_{QX}).

$$E_{QX} = \frac{[QX]}{[Q^{\oplus}] \cdot [X^{\ominus}]} \tag{2}$$

$[QX]$ = molar concentration of the ion pair in the organic phase
$[Q^{\oplus}]$ $[X^{\ominus}]$ = molar concentration of the cation resp. anion.

The magnitude of the equilibrium (i.e. magnitude of E_{QX}) is controlled by the chemical nature of the ion pair, the pH-value, by the chemical nature of the starting materials, and the organic phase.

The influence on the retention for a system with a lipophilic stationary phase and a hydrophilic mobile phase (reversed-phase system) is determined by:

$$K' = E_{QX} \cdot X^{\ominus} \cdot \frac{Vs}{Vm} \tag{3}$$

Vs = Volume of the stationary phase
Vm = Volume of the mobile phase

For a system with a hydrophilic stationary phase and lipophilic mobile phase (straight phase system) the following equation is valid:

$$K' = \frac{1}{E_{QX} \cdot [X^{\ominus}]} \cdot \frac{Vs}{Vm} \qquad (4)$$

The relationship between K' and R_f can be described by:

$$K' = \frac{1 - R_f}{R_f} \qquad (5)$$

This means, that in a reversed-phase system increasing concentrations of the counter-ion result in an increase of K', i.e. a decrease of R_f.

Suitable counter-ions for bases are sulfonic aeids and sulphates, for acids quarterny ammonium compounds.

For separation of bases the pH of the mobile phase is about 1–3.5, for separation of acidic samples about 7–9.

3 The Chemistry of Reversed-Phase Layers

According to the definition of Martin and Synge "reversed-phase" comprises a system, consisting of an apolar stationary phase and a polar mobile phase. In contrast straight phase systems are composed of a polar stationary phase and an apolar mobile phase.

In principle reversed-phase layers can be prepared in two different ways:

a) Straight phase (e.g. cellulose — or silicagel) layers are impregnated with a solvent, immiscible with water and polar solvents.
   Such solvents are
   Alcohols with a carbon chain of least at C_8
   Fatty oils (castor oil, peanut oil)
   Silicone oil
   Hydrocarbons (undecane, hexadecane, paraffin)

b) by chemical reaction with the silanol groups of the silicagel layer.
   One must distinguish between the two different types:
   1. monomeric bound phases
      These are prepared by reacting the silanol groups with trialkylchlorsilanes (Ia) or dialkylchlorsilanes (Ib) (Fig. 1)
      Reaction Ib leads to a higher degree of alkylation (because the chlor-atoms react with neighbouring silanol groups). This results in more suitable layers.
   2. polymer bound phases
      Silanol groups are reacting with di- or trifunctionel alkylsilanes in presence of small amounts of water: (Fig. 2)

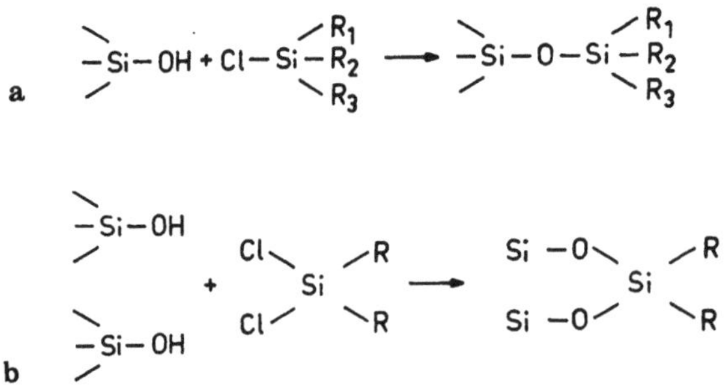

Fig. 1a and b

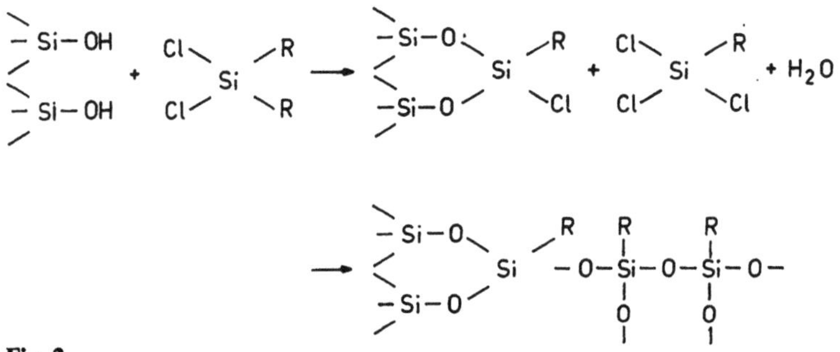

Fig. 2

Table 1. Survey of Commercial Available Chemically Bonded HPTLC-Plates

Plate designation	Manufacturer	Length of carbon chain	Comments
Plates for nano TLC			
HPTLC-precoated plates RP_2F_{254}	Merck Darmstadt, FRG	C_2	a)
HPTLC-precoated plates RP_8F_{254}	Merck Darmstadt, FRG	C_8	a)
HPTLC-precoated plates $RP_{18}F_{254}$	Merck Darmstadt, FRG	C_{18}	a)
Nano-SIL 50 UV 254	Macherey-Nagel Dueren, FRG	C_{18}	a) d)
Nano-SIL C_{18} 75 UV 254	Macherey-Nagel Dueren, FRG	C_{18}	a) e)
Nano-SIL C_{18} 100 UV 254	Macherey-Nagel Dueren, FRG	C_{18}	a)
Plates for conventionel TLC			
TLC precoated plates silanized	Merck Darmstadt, FRG	C_2	a)
TLC precoated plates RP_8F_{254s}	Merck Darmstadt, FRG	C_8	b) c)
TLC precoated plates $RP_{18}F_{254s}$	Merck Darmstadt, FRG	C_{18}	b) c)
OPTI-UP C_{12}	Antec, Benwill, Switzerland	C_{12}	b)
KC_{18}	Whatman, Clifton, (NJ) USA	C_{18}	a)
Multi K CS_5	Whatman, Clifton (NJ) USA	C_{18}	f)

a) water content limited in the mobile phase
b) water content not limited in the mobile phase
c) acid resistant fluorescence indicator
d) degree of alkylation 50%
e) degree of alkylation 75%
f) dual phase (C_{18}/silica plate)

4 Mechanism of Ion Pair Chromatography on Reversed-Phase Layers

Several authors [1-6] have offered various mechanisms regarding ion pair retention. Although these proposals concern column chromatography, the two most popular are mentioned since the conditions in thin layer chromatography are quite similar.

a) Ion pair formation occurs in the aqueous mobile phase. This ion pair is adsorbed onto the hydrophobic stationary phase. Retention is controlled by the degree of non-polarity of the ion pair. The longer the carbon chain of the pairing reagent the greater the non-polarity (lipophilie) of the ion pair.

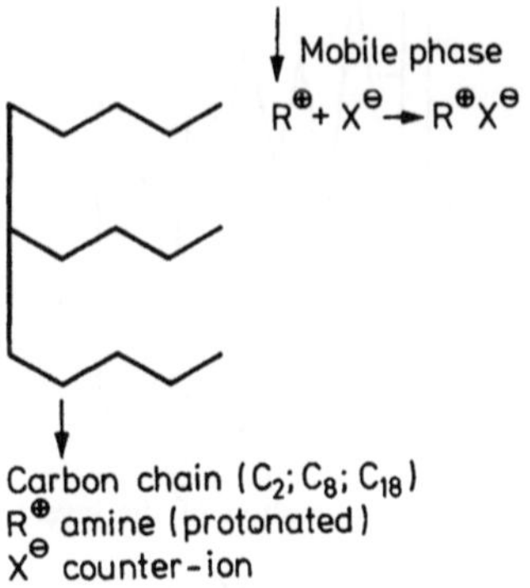

Fig. 3

b) The unpaired lipophilic ion (i.e. heptanesulfonic ion) is adsorbed onto the stationary phase. Thus the negatively loaded layer takes over the function of an ion exchanger for the cation (i.e. protonated amine).

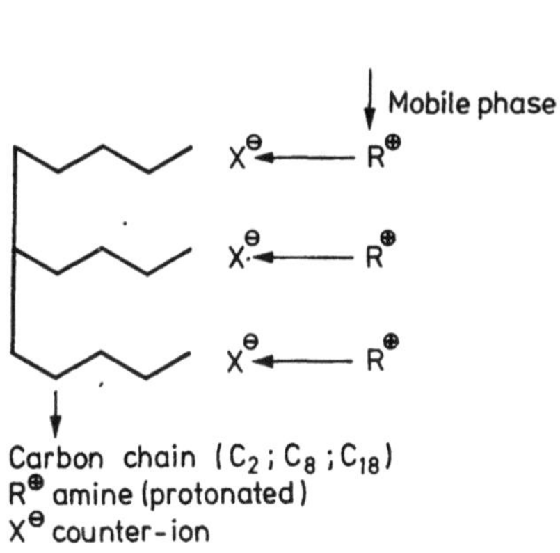

Fig. 4. Adsorption of the counter-ion on the layer

5 Technique of Ion Pair Chromatography

Volkmann [7] investigated the parameters of ion pair chromatography on HPTLC-plates. As test samples he used phenothiazine bases and the corresponding sulfoxides.

Three *techniques of performance* are possible and were tested:

1. the counter-ion (heptane sulfonic acid) was added to the mobile phase,
2. the layer was impregnated with a solution of the counter-ion,
3. both mobile phase and layer contained the counter-ion.

The results of the three techniques are shown in Fig. 5.

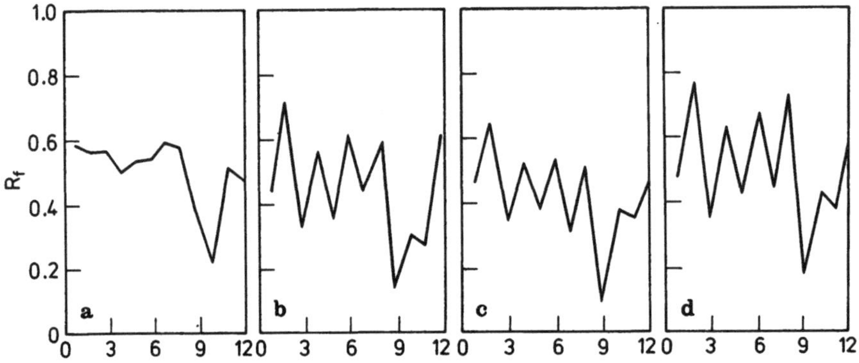

Fig. 5. **a** Without counter-ion; **b** Counter-ion in the mobile phase; **c** Impregnation of the layer; **d** Counter-ion in the mobile phase and in the layer

Sorbent: HPTLC plate precoated $RP_{18}F_{254}$ for Nano-TLC, Merck

Mobile phase: $CH_3OH:H_2O:CH_3COOH$ 80:15:5

1,3,5,7,9,11 phenothiazine base

2,4,6,8,10,12 phenothiazine sulfoxide

linear development (linear chamber Camag)

Applied volume and quantities: 100 nl = 20 ng

Detection: UV light 254 and 360 nm

A separation of phenothiazine base and sulfoxides is achieved when the counter-ion is in the mobile phase. The additional presence of the counter-ion in the stationary phase does not significantly improve separation. The presence of the counter-ion resulted not only in a complete separation but also the spots are extremely sharp.

Which of the three techniques produces the best separation depends mainly on the "chemistry" of the corresponding counter-ion. When using alkene or naphthalene-sulfonic acids it is sufficient to add the counter-ion to the mobile phase, whereas the utilization of sulfonic acid esters requires impregnation of the layer.

A further parameter is given by the *choice of the stationary phase* (Fig. 6) Separation is significantly improved by changing from RP_2 to RP_8 layers.

Concentration of the counter-ion: Compactness of spots is achieved when ion pair formation is complete. Fig. 7 shows that a concentration of 2% of the counter-ion is sufficient in this case. Increasing the concentration to 3% does not improve separation (Fig. 7).

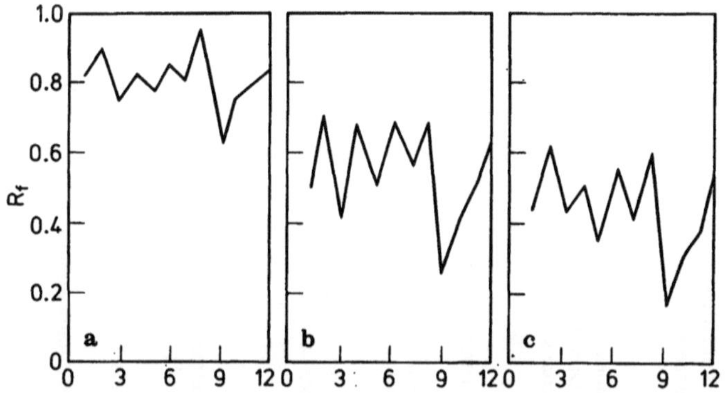

Fig. 6. a RP_2; b RP_8; c RP_{18}; other chromatographic conditions: like Fig. 5

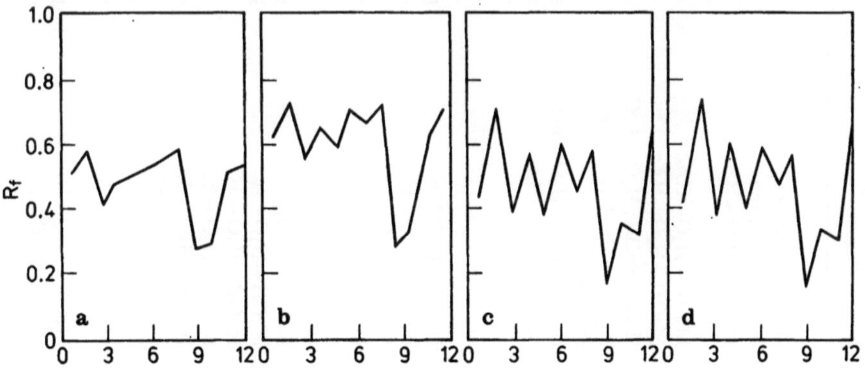

Fig. 7. Concentration of the counter-ion (in %): a 0.25; b 0.5; c 2; d 3; other chromatographic conditions: like Fig. 5

6 Literature Survey

See Table 2, page 58.

7 Chromatography on Impregnated Layers

Literature contains very few reports on ion pair chromatography upon impregnated layers. Schill et al. developed theoretically interesting systems [27–29], which allow the computation of the parameter governing ion pair chromatography. Such a system is described for the separation of certain tertiary and quaternary ammonium compounds [8]. The system uses acetylated cellulose as support, 1-butanol or 1-pentanol as stationary phase and aqueous salt solutions in sulfuric acid as mobile phase. The salt solutions, which consist of NaCl or NaBr or $NaClO_4$ and sulfuric acid play the part of counter-ions.

Table 3 lists the extraction constants of the bases used as test samples. These were determined by partition in an aqueous organic system. (E^{xx}_{HAX} = conditional extraction constant).

Table 2. Literature Survey Ion Pair Chromatography on Reversed-Phase Layers

Counter-ion	Stationary phase	Mobile Phase	Application	Ref.
1. Anionic counter-ions				
$Cl^{\ominus}$; $Br^{\ominus}$; $ClO_4^{\ominus}$	acetylated cellulose 1-butanol; 1-pentanol	aqueous solutions of the counter-ion	Ammonium compounds	8)
Heptanesulfonic acid	RP_2; RP_8; RP_{18}	MeOH—H_2O—HAc	Phenothiazine bases and sulfoxides	7)
	RP_8/RP_{18}	MeOH—H_2O—HAc	Alcaloids	9)
	SIL C_{18} 50%	MeOH—H_2O—HAc	Alcaloids anaesthetics	10)
	silanized silicagel	H_2O—MeOH; H_2O—AcCN H_2O—THF; H_2O-Acetone H_2O-dioxane	cephalosporins	11)
	silanized silicagel	H_2O—MeOH; H_2O-dioxane	sulfonamides	12)
Pentane-sulfonic-acid	RP_2; RP_8; RP_{18}	MeOH—H_2O—HAc	Phenothiazine bases and sulfoxides	13)
Octane-sulfonic-acid				
Duodecane-sulfonic-acid				
Dodecylbenzenesulphonic acid	silanized silicagel	MeOH—H_2O-HAc	primary aromatic amines	14)
	silanized silicagel	MeOH—H_2O—HAc	Dipeptides	15)
	silanized silicagel	MeOH—H_2O—Ac MeOH—H_2O—NH_3 MeOH—H_2O—HCl	Sulfonamides primary aromatic amines	16)
	silanized silicagel	MeOH—H_2O—Ac MeOH—H_2O—HCl	aliphatic monoresp. polyamines	17)
	silanized silicagel	MeOH—H_2O—Ac MeOH—H_2O + NH_3 MeOH—H_2O + buffer	phenols	18)
	silanized silicagel	MeOH—H_2O—HAc	di-, tri-, tetrapentapeptides	19)
	RP_2, RP_8, RP_{18}	MeOH—H_2O—HAc MeOH—H_2O—HAc + HCl	Amino acids Dipeptides	20)
Triethanolamine dodecylbenzenesulfonate	silanized silicagel	MeOH—H_2O—HAc	amino acids	21)
	silanized silicagel	MeOH—H_2O—HAc	primary aromatic amines	22)
	silanized silicagel	MeOH—H_2O—HAc MeOH—H_2O—HCl	dyes	23)

	silanized silicagel	MeOH—H_2O—HAc	primary aromatic amines	24)
Sodium laurylsulfate	silanized silicagel	MeOH—H_2O—HAc	amino acids	21)
		MeOH—H_2O—HCl		
	silanized silicagel	MeOH—H_2O—HAc	phenols	
		MeOH—H_2O—NH_3		
		MeOH—H_2O-buffer		
Sodium dioctylsulfosuccinate	silanized silicagel	MeOH—H_2O—HAc	amino acids	21)
		MeOH—H_2O—HCl		
	silanized silicagel	MeOH—H_2O—Ac	di-, tri-, tetrapenta-peptides	19)
2. Cationic counter-ion				
N-Dodecylpyridinium chloride	silanized silicagel	MeOH—H_2O—HAc	dyes	23)
		MeOH—H_2O—HCl		
	silanized silicagel	MeOH—H_2O—HAc	dipeptides	15)
	silanized silicagel	MeOH—H_2O—HAc	primary	16)
		MeOH—H_2O—NH_3	aromatic amines	
		MeOH—H_2O—HCl	sulfonamides	
	silanized silicagel	MeOH—H_2O—HAc	aliphatic mono-	17)
		MeOH—H_2O—HCl	and polyamines	
Tetrabutylammoniahydroxide	silanized silicagel	MeOH—H_2O; AcCN—H_2O	Cephalosporins	11)
		THF—H_2O; Acetone—H_2O		
		dioxane—H_2O		
Tetrabutylhydrogensulfate	KC_{18}	MeOH—AcCN—THF—H_2O	Benz[α]pyrenemetabolites	25)
Tetraalkylats	RP_2; RP_8; RP_{18}	MeOH—H_2O	Dyes	26)

Table 3.

Sample	aqueous phase	organic phase	log E_{HAX}^{xx}
Papaverine	NaCl 0.1 M H_2SO_4 0.5 M	1-butanol	0.68
	NaBr 0.1 M H_2SO_4 0.5 M	1-pentanol	1.03
	$NaClO_4$ 0.1 M H_2SO_4 0.5 M	1-pentanol	1.74
Secergan®	NaCl 0.1 M H_2SO_4 0.5 M	1-pentanol	0.67
Strychnine	NaCl 0.1 M H_2SO_4 0.5 M	1-pentanol	—0.13

The chromatographic systems allow the determination of Vs (volume of the stationary phase) and Vm (volume of the mobile phase). This again allows the computation of the optimal migration distance using the following equation:

$$\frac{1 - R_f}{R_f} = E_{HAX}^{xx} \cdot Cx \cdot \frac{Vs}{Vm} + E_{HAY}^{xx} \cdot Cy \cdot \frac{Vs}{Vm} \tag{6}$$

If the concentration of Cy (e.g. $HSO_4^{\ominus}$) is kept constant, a linear relationship between $(1 - R_f)/R_f$ and Cx (e.g. $Cl^{\ominus}$) exists. (Fig. 8)

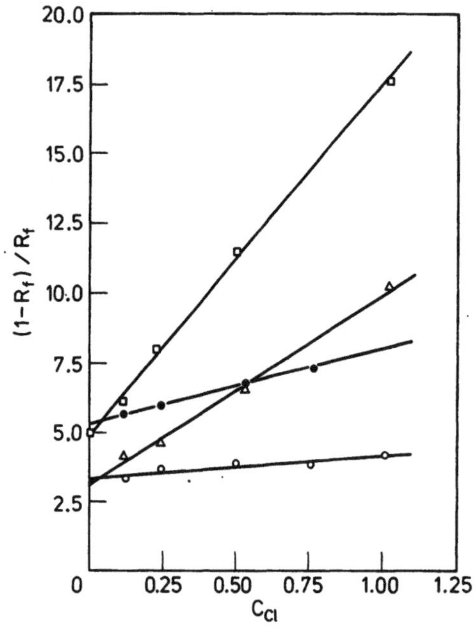

Fig. 8. Strychnine as ion pair with chloride on different stationary phases
Support: Cellulose with 10% acetyl content
Mobile phase: NaCl in 0.5 M H_2SO_4
Stationary phase: 0.5 g/g of support
1-butanol = □; 1-pentanol = △; 1-hexanol = ○; 1-octanol = ●

The slope (E_r) and the intercept (Int) of the straight line may be calculated as follows:

$$E_r = E^{xx}_{HAX} \cdot \frac{Vs}{Vm} \tag{7}$$

$$\text{resp. Int} = E^{xx}_{HAHSO_4^{\ominus}} \cdot C_{HSO_4^{\ominus}} \cdot \frac{Vs}{Vm} \tag{8}$$

These two constants (E_r and Int) allow the computation of the optimal counter-ion concentration Cx (Cy kept constant) necessary to separate a base pair.

$$Cx = \frac{E_{r1}^{1/2} \cdot (1 + Int_2) - E_{r2}^{1/2} \cdot (1 + Int_1)}{(E_{r1} \cdot E_{r2})^{1/2} \cdot (E_{r1}^{1/2} - E_{r2}^{1/2})} \tag{9}$$

Cx inserted in Eq. (6) yields the R_f-values for the base pairs (e.g. for the separation of papaverine and strychnine).

Table 4.

Counter-ion in the mobile phase	stationary phase	Cx	R_f	
			Papaverine	Strychnine
NaCl	1-butanol	0.93	0.27	0.37
NaCl	1-pentanol	0.70	0.28	0.56
NaCl	1-hexanol	1.30	0.28	0.62
NaBr	1-pentanol	0.26	0.37	0.61

The advantage of the system, which can also be used for straight phase systems, is its great flexibility. This is particularly valid for reversed-phase systems as retention of the ion pair is clearly controlled by the concentration of the ion pair in the mobile phase. The system makes it possible to determine Vs and Vm. In this way retention factor K′ can be calculated as follows:

$$K' = E_{QX} \cdot [X^{\ominus}] \cdot \frac{Vs}{Vm} \tag{10}$$

Consequently it is possible to transfer data from thin layer chromatography to column chromatography.

The relationship:

$$\frac{1 - R_f}{R_f} = E_r \cdot Cx + Int \tag{11}$$

yields R_f-values which are higher than these found by chromatography. The reasons for this are the adsorptive properties of the support.

Table 5.

Sample	counter-ion	stationary phase	R_f (Cx = 0.1)	
			calc.	found
Papaverine	NaCl	1-butanol	0.69	0.59
	NaCl	1-pentanol	0.74	0.56
	NaBr	1-pentanol	0.66	0.46
		1-pentanol	0.35	0.12

8 Chromatography on Chemically Modified Layers

8.1 Chromatography on Silanized Silicagel

Papers on ion pair chromatography on silanized silicagel layers are published by Lepri et al. [14–19,21–24]. Anionic and cationic detergents were used as counter-ions. These were introduced into the system by layer impregnation.

The system can be modified using the following parameters:

1. kind of counter-ion
2. concentration of the counter-ion
3. pH of the mobile phase
4. ratio methanol: water

The aim of all these parameters is to influence the extraction constant (E_{QX} value). In most cases this alteration results in an increase or decrease of retention, in some cases in a change in selectivity, i.e. a change in the sequence of the samples on the chromatogram.

These effects are demonstrated by data presented in the papers of Lepri:

Table 6. Dependence of Retention on the Concentration of the Counter-ion [24]
Adsorbent: silanized silicagel "Merck"
Mobile phase: $H_2O—CH_3OH—CH_3COOH$: 54.3:40:5.7
Counter-ion: dodecylbenzenesulfonate
Concentration of the counter-ion: a) 0 b) 0.25 c) 0.5 d) 1.0 e) 2.0 f) 3.0 g) 4.0
(in %)

Sample	R_f-values						
	a	b	c	d	e	f	g
Octopamine	0.92	0.81	0.80	0.63	0.43	0.37	0.38
Histamine	0.90	0.72	0.60	0.36	0.15	0.10	0.07
Tryptamine	0.67	0.49	0.38	0.22	0.15	0.10	0.09
Tyramine	0.80	0.71	0.62	0.48	0.35	0.26	0.26
Noradrenaline	0.83	0.83	0.78	0.09	0.54	0.46	0.48

These examples demonstrate the flexibility of the developed system. Some points, however, require a more detailed explanation.

a) *Increase of counter-ion concentration*: The use of counter-ions with long carbon chains (e.g. C_{12}) forms hydrophobic ion pairs. This results in increased retention on reversed-phase layers.

However, when using hydrophilic counter-ions (e.g. Hydroxynaphthalene, sulfonic acids) [30] a decrease in retention is observed, because the ion pair becomes more hydrophilic in respect to the uncomplexed amine. Furthermore increased counter-ion concentration as well as a pH change prevent the dissociation of the ion pair.

Table 7. Dependence on the Kind of Counter-ion [21]

Adsorbent:	silanized silicagel "Merck"
Mobile phase:	H_2O—CH_3OH—CH_3COOH: 64.3:5.7:30
Concentration of the counter-ion:	4%
Counter-ion:	a) without b) sodium laurylethero-sulphate c) sodium dioctylsulfo-succinate d) dodecylbenzene-sulfonic acid

Sample	R_f-values			
	a	b	c	d
Gly	0.96	0.96	0.94	0.54
Ala	0.96	0.96	0.94	0.52
β-Ala	0.96	0.73	0.47	0.35
Met	0.95	0.85	0.63	0.21
Lys	0.96	0.56	0.14	0.06
His	0.96	0.53	0.13	0.06

Table 8. Dependence of the pH-Value on the Mobile Phase [21]

Adsorbent:	silanized silicagel "Merck"
Mobile phases:	1) HCl + CH_3COOH in CH_3OH/H_2O 2) NaCl + CH_3COOH in CH_3OH/H_2O 3) CH_3COONa + CH_3COOH in CH_3OH/H_2O
Counter-ion:	dodecylbenzenesulfonic acid
Concentration of the counter-ion:	4%
pH-values in the mobile phases:	a) 0.7 b) 1.25 c) 2.75 d) 4.10 e) 5.10 f) 6.10

Sample	R_f-values					
	a	b	c	d	e	f
Gly	0.83	0.70	0.72	0.76	0.76	0.83
Ala	0.74	0.64	0.68	0.74	0.77	0.82
Val	0.54	0.33	0.44	0.60	0.67	0.74
Tyr	0.45	0.34	0.44	0.60	0.57	0.76
Glu	0.83	0.76	0.76	0.80	0.85	0.90
Lys	0.47	0.14	0.24	0.26	0.27	0.76

Table 9. Dependence of the R_f-Values on the Concentration of the Counter-ion (Without Formation of an Ion Pair) [16]

Adsorbent: silanized silicagel "Merck"

Mobile phases:
a) 0.1 M CH_3COOH + 0.1 M NH_3 in water/methanol (30%)
b) 0.1 M CH_3COOH + 0.1 M NH_3 in water/methanol (20%) pH 9.2
c) 0.1 M CH_3COOH + 0.1 M NaCl in water/methanol (20%) pH 3.25
d) 0.1 M HCl + 0.1 M CH_3COOH in water/methanol (20%) pH 1.40
e) 0.1 M CH_3COOH in water/methanol (20%)

Counter-ion: N-dodecylpyridiniumchloride

Concentration of the counter-ion: a) 0% b) 4% c) 4% d) 4%

Sample	R_f-values				
	a	b	c	d	e
Aniline	0.54	0.20	0.22	0.71	0.75
p-Toluidine	0.35	0.18	0.14	0.68	0.71
m-Phenylendiamine	0.73	0.52	0.49	0.94	0.96
o-Phenylendiamine	0.67	0.41	0.37	0.78	0.80
p-Phenylendiamine	0.75	0.67	0.60	0.94	0.98
α-Naphtylamine	0.21	0.07	0.03	0.06	0.30

Table 10. [15]

Adsorbent: silanized silicagel

Mobile phases:
a) 0.1 M CH_3COOH + 0.1 M CH_3COONa in methanol/water (30%)
b) 1 M CH_3COOH in methanol/water (30%)
c) like a
d) 1 M CH_3COONa in methanol/water (30%)

Counter-ion: N-dodecylpyridiniumchloride

Concentration of the counter-ion: a) 0% b) 4% c) 4% d) 4%

Sample	R_f-values			
	a	b	c	d
Gly-Tyr	0.96	0.97	0.96	0.96
Gly-Val	0.96	0.97	0.95	0.93
Gly-Met	0.96	0.96	0.90	0.89
Gly-Trp	0.75	0.68	0.39	0.32
Gly-Gly	0.96	0.97	0.96	0.96
Gly-His	0.96	0.97	0.96	0.81

b) *Change of retention without ion pair formation*: If a counter-ion with the same charge as the sample is added, no ion pair formation takes place. In the chromatogram a decrease of retention is observed. The two reasons have already been mentioned by Lepri:

1. an interaction between the amine (or acid) and the carbon chain of the detergent;
2. a repulse of the two equal loads (e.g. protonated base and cationic detergent).

c) *Variation of methanol-water ratio*: An increased concentration of methanol results in a decrease of retention, because of the increasing solubility of the ion pair in the mobile phase.

8.2 Chromatography on RP_2-, RP_8- and RP_{18} Layers

Another way of changing selectivity is to use plates with different long carbon chains (c.f. Technique of ion pair chromatography, chapter 5). Lepri et al. [20], T. Okumura [12] and Volkmann [7,9,10,13] investigated the availability of this material. Lepri separated amino acids and dipeptides, the layers were impregnated with dodecylbenzenesulfonic acid.

Table 11. [20]

Adsorbent:	RP_2; RP_8; RP_{18} HPTLC precoated plates "Merck"
Mobile phase:	1 M CH_3COOH in methanol/water (50%)
Counter-ion:	dodecylbenzenesulfonic acid
Concentration of the counter-ion:	4%

Sample	R_f-values		
	RP_2	RP_8	RP_{18}
Glu	0.40	0.26	0.25
Ser	0.38	0.24	0.21
Thy	0.37	0.21	0.19
Ala	0.32	0.18	0.14
Gly-Glu	0.40	0.26	0.24
Gly-Thr	0.38	0.22	0.20
Gly-Phe	0.12	0.03	0.03

T. Okumura and T. Nagaoka [12] separated sulfonamides on silanized silicagel, C_{12}- and C_{18}-layers. They used heptane sulfonic acid as a counter-ion, which was dissolved in the mobile phase. The pH of the mobile phase was not published. There was no further acid except that of the counter-ion in the mobile phase. Hence the pH should be slightly acid. The $H^{\oplus}$-concentration, however, is not sufficient to convert the sulfonamides to cations.

This, however, is necessary to form an ion pair. For ion pair separation of sulfonamides quarterny ammonia bases such as tetrabutylammonia hydroxide are suitable counterions at pH 8–9. Thus it is doubtfull whether an ion pair separation actually took place.

M. Marshall et al. [25] used tetrabutylammoniumhydrogen sulfate and mixtures of water-acetonitrile and tetrahydrofurane to separate the main metabolites and conjugates of benzo[a]pyrene on KC_{18}-plates (Whatman). In his opinion reversed-phase TLC is a very efficient method, however, it does not have the separation power of HPLC.

Table 12. [11)]

Mobile phases: a) $CH_3OH—H_2O$ 1:1
b) +0.005 M heptane sulfonic acid
c) $CH_3OH—H_2O$ 1:2 + 0.005 M heptane sulfonic acid

Sample	R_f-values			
	a	b	c	d
7-Aminocephalosporanic acid	0.93	0.89	0.92	0.85
Cephaloglycin	0.81	0.71	0.64	0.56
Cephalexin	0.81	0.70	0.71	0.65
Cephaloridine	0.68	0.58	0.51	0.48
Cephalothin	0.83	0.75	0.76	0.57

Table 13. [11)]

Mobile phases: a) $CH_3OH:H_2O$ 1:1
b) +0.005 M Tetrabutylammoniumphosphate
c) $CH_3OH:H_2O$ 1:2
d) +0.005 M Tetrabutylammoniumphosphate

Sample	R_f-values			
	a	b	c	d
7-Aminocephalosporanic acid	0.93	0.81	0.92	0.80
Cephaloglycine	0.81	0.78	0.64	0.50
Cephalexin	0.79	0.61	0.64	0.29
Cephaloridine	0.60	0.46	0.32	0.10
Cephalothin	0.78	0.59	0.49	0.22

Table 14. [9)]

Adsorbent: $RP_{18}F_{254s}$ Merck (Art. No.: 15388)
Counter-ion: sodium heptanesulfonate
Concentration: 2%
Mobile phases: a) $CH_3OH:H_2O:CH_3COOH$ 5:5:0.1
b) $CH_3OH:H_2O:CH_3COOH$ 6:4:0.1
c) $CH_3OH:H_2O:CH_3COOH$ 7:3:0.1
d) $CH_3OH:H_2O:CH_3COOH$ 8:2:0.1

Sample	R_f-values			
	a	b	c	d
Atropine	0.14	0.23	0.50	0.52
Scopolamine	0.20	0.30	0.57	0.57
Quinine	0.03	0.06	0.24	0.29
Strychnine	0.06	0.13	0.23	0.32
Brucine	0.10	0.18	0.37	0.38
Codeine	0.23	0.39	0.58	0.67
Narcotine	0.06	0.17	0.37	0.43
Papaverine	0.05	0.15	0.28	0.33

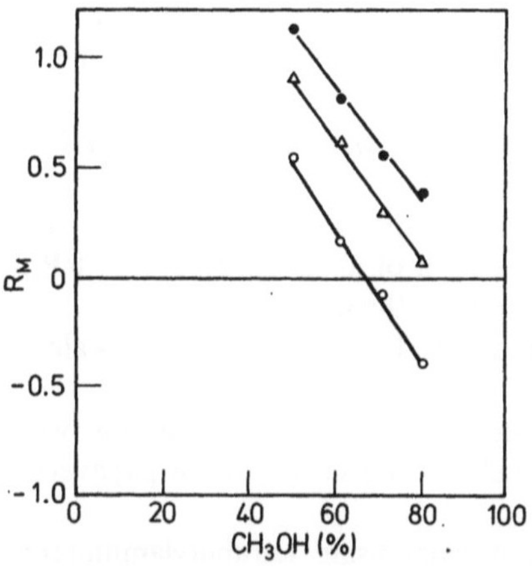

Fig. 9. Correlation between R_M value and water content of the mobile phase

Adsorbent: TLC precoated plates $RP_{18}F_{254s}$ Merck, Darmstadt, FRG No. 15423

Mobile phase: $CH_3OH:H_2O:CH_3COOH$

Counter-ion: heptane sulfonic acid 2%

● Strychnine; × Brucine; ○ Codeine

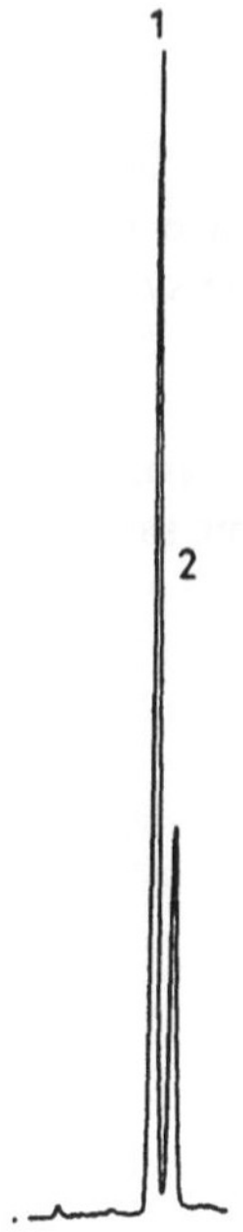

Fig. 10. 1. Quinine; 2. Papaverine

Adsorbent:	SIL C_{18} 50% Machery-Nagel, Dueren, FRG
Mobile phase:	$CH_3OH:H_2O$ $:CH_3COOH$ 5:5:0.5 + 0.2 g heptane sulfonic acid
Circular development:	Flow rate: 2 sec μl^{-1}
Fluorescence measurement:	Zeiß chromatogram Scanner KM_3

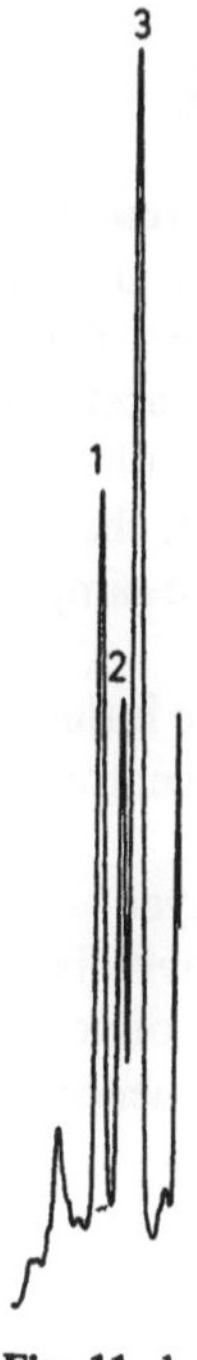

Fig. 11. 1. Evipan®; 2. Luminal®; 3. Veronal®

Sorbent:	SIL C_{18} 50% Macherey-Nagel, Dueren, FRG
Mobile phase:	$CH_3OH:H_2O$ 5:5 + 1 ml 0.1 N TBAH
Remission measurement:	Zeiß-Chromatogram Scanner KM_3

T. Okumura used heptane sulfonic acid and tetrabutylammonia phosphate for the separation of cephalosporin antibiotics. The addition of ion pair formers results in increased retention without any change in sequence. [11]

These results obtained with silanized silicagel permit a direct correlation to HPLC data on RP_2 columns.

Volkmann applied ion pair chromatography to the separation of alkaloids, anaesthetics and barbiturates [9,10]. For the separation of alkaloids RP_8- and RP_{18}-layers were used which allow a higher water content in the mobile phase.

The two systems used showed a linear relationship between the R_M of alkaloids and the water content in the mobile phase.

Volkmann also used partially alkylated layers [10]. The chromatograms were developed in the U-chamber (Camag). Fig. 10 shows the separation of papaverine and quinine.

Likewise barbiturates were separated as ion pairs using tetrabutylammonium hydroxide. (Fig. 11).

9 Summary and Outlook

Ion pair chromatography represents a valuable variant in thin layer chromatography. Reversed-phase layers are particular suitable as these layers have no or only a few adsorptive properties. Chromatography is possible with simple systems, e.g. mixtures of methanol or acetonitrile with water. In contrast to HPLC it is possible to use mobile phases with $pH > 8$.

The addition of a counter-ion changes the selectivity. Thus not only the separation of ionogenic and non ionogenic samples is improved, but also the ionogenic samples themselves.

Changes of selectivity can be influenced by:

a) chemical structure and concentration of the counter-ion
b) pH of the mobile phase
c) chemical structure of the layer

The result of changed selectivity is:

a) increased retention (in case of very hydrophilic counter-ions a decrease)
b) in some cases change of sequence
c) increase of spot compactness

The aims of future investigations in ion pair chromatography should be

1. to find other suitable counter-ions. These should have
   a) a strong influence on selectivity
   b) chemical properties to improve detection limit (i.e. to produce fluorescence)
2. to find out the best techniques for the various reversed-phase layers;
3. to investigate the role of the reversed-phase layer on the separation

10 References

1. Wittmer, D. P., Nuessle, N. O., Haney, W. G.: J. Anal. Chem. *47*, 1422 (1975)
2. Horvath, C., Melander, W., Molnar, I., Molnar, P.: ibid. *49*, 2295 (1977)
3. Horvath, C., Melander, W., Molnar, I.: J. Chromatogr. *125*, 129 (1976)

4. Venne, J. L. M., Hendrix, J. L. H. M., Decdler, R. S.: ibid. *167*, 1 (1978)
5. Kraak, J. C., Jonker, K. M., Huber, J. F. K.: ibid. *142*, 671 (1977)
6. Kissinger, P. T.: J. Anal. Chem. *49*, 883 (1977)
7. Volkmann, D.: HRC u. CC *2*, 729 (1979)
8. Gröningsson, K.: Acta Pharm. Suecica *7*, 635 (1970)
9. Volkmann, D.: HRC u. CC *4*, 350 (1981)
10. Volkmann, D.: in "Chromatographic Methods; Instrumental HPTLC" (W. Bertsch, S. Hara, R. E. Kaiser, A. Zlatkis, Hrsg.), Huethig Verlag, Heidelberg 1975
11. Okumura, T.: J. Liquid Chromatogr. *4*, 1035 (1981)
12. Okumura, T., Nagaoka, T.: ibid. *3*, 1947 (1980)
13. Volkmann, D.: Kontakte (Merck) *1981*, (3) 32
14. Lepri, L., Desideri, P. G., Heimler, D.: J. Chromatogr. *153*, 77 (1978)
15. Lepri, L., Desideri, P. G., Heimler, D.: ibid. *207*, 412 (1981)
16. Lepri, L., Desideri, P. G., Heimler, D.: ibid. *169*, 271 (1979)
17. Lepri, L., Desideri, P. G., Heimler, D.: ibid. *173*, 119 (1979)
18. Lepri, L., Desideri, P. G., Heimler, D.: ibid. *195*, 339 (1980)
19. Lepri, L., Desideri, P. G., Heimler, D.: ibid. *195*, 187 (1980)
20. Lepri, L., Desideri, P. G., Heimler, D.: ibid. *209*, 312 (1981)
21. Lepri, L., Desideri, P. G., Heimler, D.: ibid. *195*, 65 (1980)
22. Lepri, L., Desideri, P. G., Heimler, D.: ibid. *155*, 119 (1978)
23. Lepri, L., Desideri, P. G., Heimler, D.: ibid. *161*, 279 (1978)
24. Lepri, L., Desideri, P. G., Heimler, D.: ibid. *153*, 77 (1978)
25. Marshall, M. V., Gonzales, M. A., McLemore, T., Busbee, D. L., Wray, N. P., Griffin, A. C.: ibid. *197*, 217 (1980)
26. Gonnet, C., Marichy, M., Naghizadey, H.: Recent. Develop. Chrom. Elektroph. *10*, 11 (1980)
27. Gröningsson, K., Schill, G.: Acta Pharm. Suecica *6*, 447 (1969)
28. Gröningsson, K., Weimers, M.: ibid. *12*, 65 (1975)
29. Gröningsson, K., Westerlind, H., Modin, R.: ibid. *12*, 97 (1975)
30. Volkmann, D.: HRCuCC in preparation

Evaluation and Calibration in Quantitative Thin-Layer Chromatography

Siegfried Ebel

Institut für Pharmazie und Lebensmittelchemie, Universität Würzburg, Am Hubland, 8700 Würzburg, FRG

Table of Contents

1 Basic Principles . . . 72
1.1 Kubelka-Munk-Theory . . . 72
1.2 Problems Caused by Substance Inhomogenity . . . 74

2 Evaluation . . . 76
2.1 Peak height . . . 76
2.2 Integrated Area . . . 77
2.3 Spline Interpolation . . . 79
2.4 Derivative Recording . . . 80
2.5 Peak Approximation . . . 81

3 Calibration . . . 82
3.1 Linear Regression . . . 82
3.2 Regression with Linearization . . . 84
3.3 Non-linear Regression . . . 84

4 Standard Techniques . . . 88
4.1 External Standard Method . . . 89
4.2 Internal Standard Method . . . 90
4.3 Standard Addition Method . . . 91

5 Limit of Detection and Determination . . . 91
5.1 Limit of Detection . . . 91
5.2 Limit of Determination . . . 91

6 Acknowledgements . . . 92

7 References . . . 93

1 Basic Principles

In addition to transfer techniques in quantitative analyses, such as polarography, titrimetry, spectroscopy and other analytical methods used after separation by TLC, the *in situ* optical measurement is the most widely employed technique for quantitative determinations. In most cases UV-absorption is used, while coloured substances can be determined by absorption measurement in the visible range of the spectrum. Fluorescent substances are preferably determined by fluorescence measurement. Infrared absorption techniques are not used in routine analysis up to this date.

In order to understand the problems arising in quantitative analyses employing in situ-methods, we have the look at what happens with an incident light beam within the sorbens layer. On the other hand, we have to investigate the problems caused by the inhomogenity of the concentration of the substance within a single spot.

1.1 Kubelka-Munk-Theory

In 1760, Lambert observed that the white wall of a house bathed in sunlight has the same brightness from whatever angle it is viewed. Starting from this observation, he derived the first law of diffuse reflection or scattering, the basis of all modern models of reflectance theory. In an ideal scattering medium we can rationalize all fluxes of light as components of two vectors: I_T, in the direction of the incident light, and J_R, antiparallel thereto. Owing to absorption (symbolized by k) and scattering (symbolized by s) the intensity of these two fluxes are diminished in a manner described by the differential Eqs. (1) and (2) [1,2] (cf. Fig. 1). At first glance, these two equations appear to be very simple; however, as yet there is no explicit solution. The first approximation valid for an almost infinite scattering medium, such as the atmosphere, was undertaken in the field of astrophysics [3].

$$dI_T = -2(k + s)\, I_T\, dx + 2sJ_R\, dx \tag{1}$$

$$dJ_R = -2(k + s)\, J_R\, dx - 2sI_T\, dx \tag{2}$$

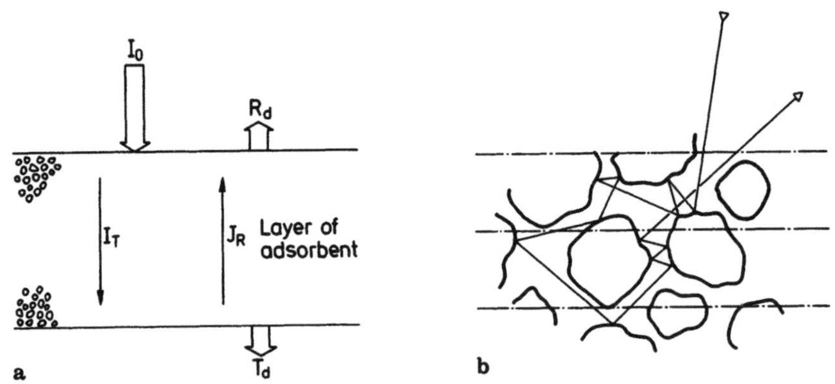

Fig. 1. Interaction of adsorbent layer with incident light [4].
a Schematic fate in scattering; **b** Fluxes of incident, transmitted and diffuse reflected light

The most convenient approximate solutions have been given by Kubelka and Munk [5,6] (cf. [1]). These are valid only in the case of an ideal scattering medium without any regular reflection. While the so-called linear Eq. (3) is only valid at infinite thickness, the so-called hyperbolic solution (4) and (5) is of the greatest utility in TLC and HPTLC with a finite layer thickness d. It must be mentioned that R_∞ in Eq. (3) is the absolute reflectance, while in optical in situ reflectance measurements in TLC and HPTLC only relative reflectances are accessible.

$$F_{KM} = \frac{(1 - R_\infty)^2}{2R_\infty} = \frac{k}{s} \tag{3}$$

$$R_d = \frac{\sigma \cdot \sinh(d \cdot q)}{(\alpha + \varrho) \sinh(d \cdot q) + q \cdot \cosh(d \cdot q)} \tag{4}$$

$$T_d = \frac{q}{(\alpha + \varrho) \sinh(d \cdot q) + q \cdot \cosh(d \cdot q)} \tag{5}$$

α absorptivity

ϱ coefficient of scattering

$$q = \sqrt{\alpha^2 + 2\alpha \cdot \sigma} \tag{5a}$$

The equations derived by Bodó [7] (6) and (7) are seen to be similar to (4) and (5) if we consider the definitions of the hyperbolic functions (8) and (9).

$$R_d = \frac{\varrho \cdot e^{\alpha d} + \varrho(1 - 2\varrho) e^{-\alpha \cdot d}}{e^{\alpha \cdot d} - \varrho^2 e^{-\alpha \cdot d}} \tag{6}$$

$$T_d = \frac{1 - \varrho^2}{e^{\alpha \cdot d} - \varrho^2 \cdot e^{-\alpha \cdot d}} \tag{7}$$

$$\sinh(x) = \frac{1}{2}(e^x - e^{-x}) \tag{8}$$

$$\cosh(x) = \frac{1}{2}(e^x + e^{-x}) \tag{9}$$

Theoretical consideration of these problems in scattering has important aims: to reach an understanding of the measuring principle itself and to help the analyst in finding more precise methods of evaluating the data. One way to examine the fate of the incident light is to divide the layer in these subdivisions and to calculate transmittance, absorbance and scattering in each sublayer. Such multilayer models have been discussed by Post [8], Prošek [9,10] and Ebel and Post [4].

Some typical results are presented in Figs. 2–4. The reflectance in a typical wide range calibration is shown in Fig. 2. The similarity to the hyperbolic Kubelka-Munk function is obvious. The theory predicts an increase in reflectance with increasing layer thickness or decreasing particle size. The relationship between the squared reflectance R_d^2 and the amount of substance per spot (Fig. 3) is almost linear, as found empirically by Tausch [11]. At low amounts of substance per spot there is a linear relationship (Fig. 4) first published by Huber [12] on a special case of a non-linear equation and later on by Frei [13] and by Ebel and coworkers [14,15].

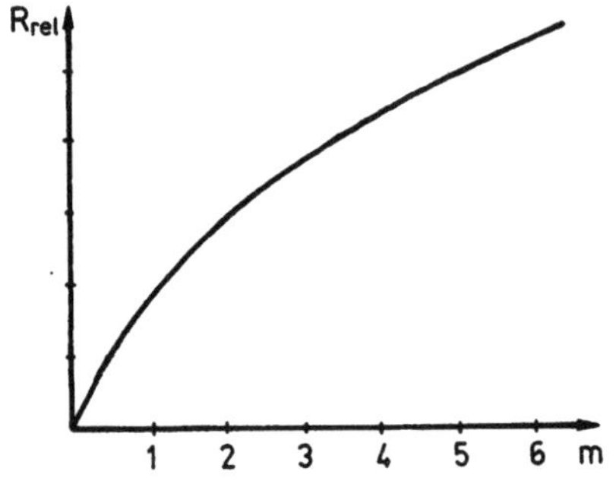

Fig. 2. Calculated relationship between reflectance and amount of substance per spot(m) in a wide-range calibration by a multilayer modell [4]

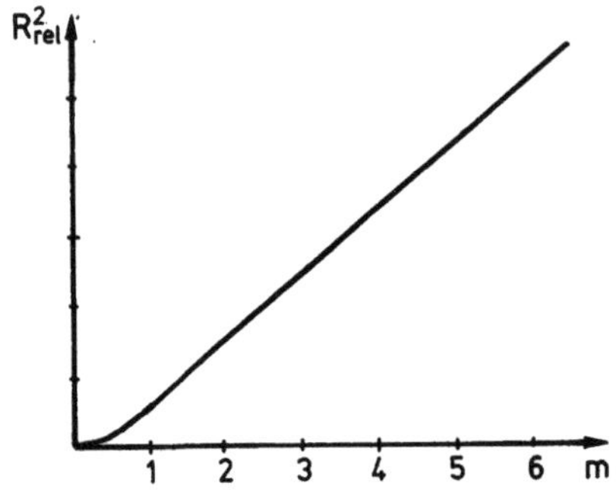

Fig. 3. Calculated squared reflectance by multilayer model according to Tausch [11] in a wide-range calibration

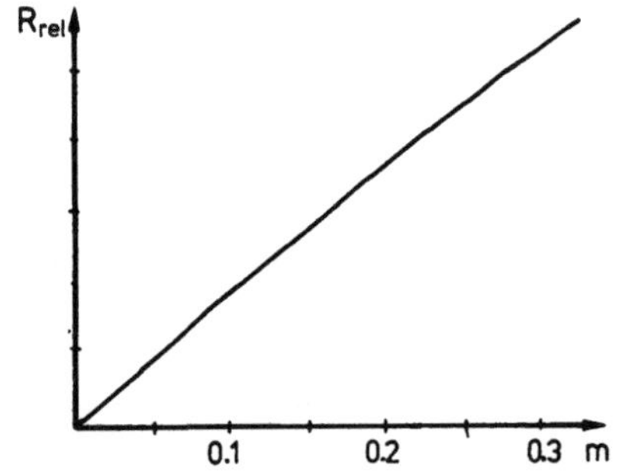

Fig. 4. Calculated approximatly linear relationship of reflectance vs. amount per spot by multilayer model [4] in a small-range calibration

1.2 Problems Caused by Substance Inhomogenity

Using spectrophotometric techniques in homogenous solutions, the primary measured signal is given by the transmittance T. From this absorbance is calculated by (10). The absorbance A is strongly related to the concentration c by the absorptivity α. This relationship is the well-known Boguert-Lambert-Beer-law (11).

$$A = \ln \frac{I_0}{I_m} = -\ln T \tag{10}$$

$$A = \alpha \cdot c \cdot d \tag{11}$$

I_0 intensity of the incident light
I_m intensity to be measured
T transmittance
A absorbance
α absorptivity
c concentration
d depth of the cell used

While in the case of transmittance measurements the light beam penetrates homogeneous solutions, reflectance in TLC involves inhomogeneous concentrations in the spot and, therefore, within the incident light beam. Caused by the chromatographic process the concentration profile within the spot is given from the binomial distribution valid in TLC [16,17,18] by Eq. (12). It is noteworthy that the gaussian distribution will lead to a wrong Eq. (13) containing the term R_f instead of $R_f(1 - R_f)$. The binomial distribution will lead to narrow peak near the front, while the gaussian distribution gives broadened peaks also beyond the solvent front. All equations contain only the concentration profile in the direction of development y.

$$c = \frac{c_{max}}{\sqrt{2\pi \cdot N \cdot R_f(1 - R_f)}} \cdot \exp\left\{-\frac{(y - y_{max})^2}{2N \cdot R_f(1 - R_f)}\right\} \tag{12}$$

$$c = \frac{c_{max}}{\sqrt{2\pi \cdot N \cdot R_f}} \cdot \exp\left\{-\frac{(y - y_{max})^2}{2N \cdot R_f}\right\} \tag{13}$$

N number of theoretical plates

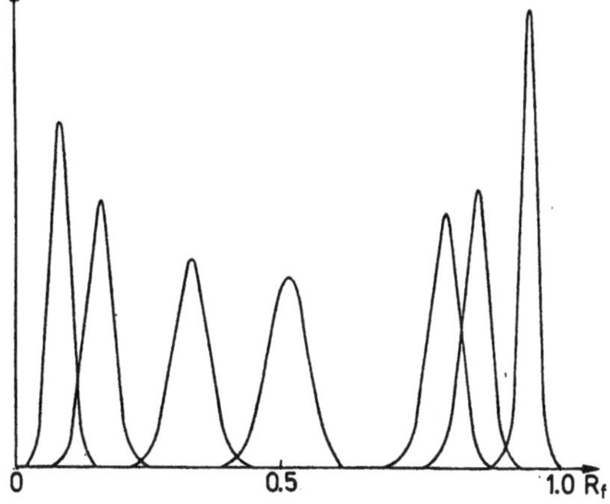

Fig. 5. Concentration profiles of TLC-spots with different R_f-values given by a binomial distribution without broadening by diffusion [18]

At each point within the spot there is a different concentration which will give an absorption of light and a nonlinear reflectance according to the Kubelka-Munk-law. The photomultiplier is now integrating all the remitted light and, therefore, a very complex and not easily calculable mean signal is produced. From the Kubelka-Munk-law and from this mean signal of an inhomogeneous reflectance it should be understood that there cannot be a linear relationship between the measured signal and the amount of substance to be determined on a specific spot.

2 Evaluation

The scanning of a track of a developed plate by a slit-scanner yields the space dependent measuring curve of the reflectance (Fig. 6). From this curve data must be taken for calibration or for calculating the analytical result. This may be the peak-height or the integrated area under the curve. We also may produce the derivative of this curve to obtain data, or in the case of a digital system we may obtain data for approximation of the peak.

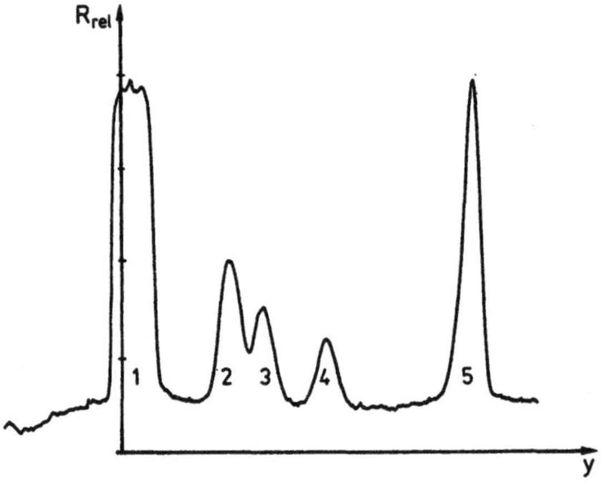

Fig. 6. Typical recorded space dependent measuring curve in reflectance mode.
Substances: Aglykones of an extract from the root of Rheum officinale [19]
1. Position of start with the non chromatographed matrix and the glycosides; 2. rhein; 3. physcione; 4. emodine; 5. chrysophanol

2.1 Peak height

In case no complications arise, as e.g. through tailing it is possible to determine the maximum height of the recorded peak and take only this one data for calibration or calculating the analytical result. The advantage of measuring peak-height is the high reproducibility and precision. On the other hand, there is a disadvantage if there are nonseparated peaks, tailing peaks or small changes of R_f-values within the same plate. From our experience according to the literature, peakheight will lead to better results than integrated areas in all cases where peakheight measurement is possible and not falsified by chromatography or interfering substances. Peakheight has another advantage compared with integrated areas: if complex mixtures of substances are to be separated as in the analysis of naturally occurring drugs or of drugs and their metabolites in body fluids the separation will often show a similar result as demonstrated in Fig. 7. The well but not perfectly separated peaks of rhein an physcione slightly overlap, but the determination of both peakheights is not disturbed by the other (baseline definition!).

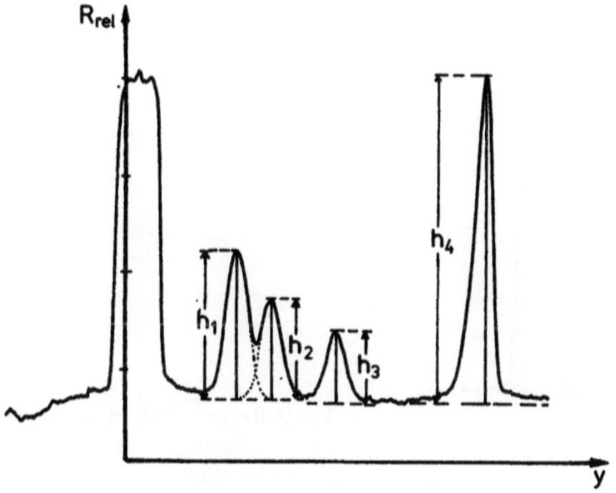

Fig. 7. Recorded reflectance curve with evaluation of peakheight. Peakheight of not ideally separated peaks is only slightly disturbed by the neighbouring peak

2.2 Integrated Area

Integration of peak areas in routine analysis is mostly achieved with electronic integrators. Two main problems arise from this evaluation technique: first the definition of the baseline, and second, the determination of the time constant of the TLC-scanner and the integrator electronics.

Most integrators are developed for use in GC and have sophisticated routines for peak detection and for baseline corrections if there are negative slopes. They are able to detect small peaks even in the descending part of a solvent peak. How-

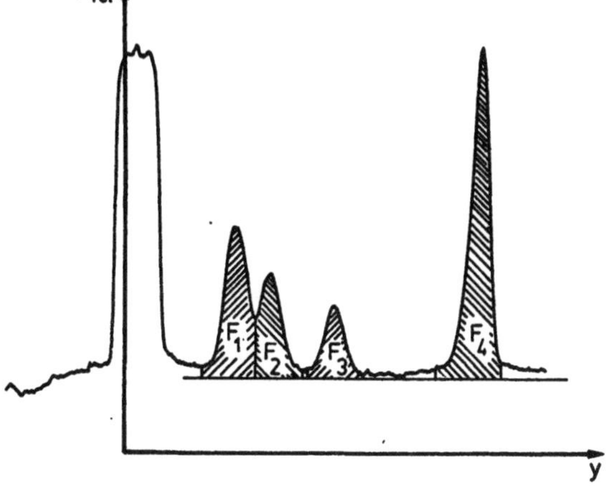

Fig. 8. Recorded reflectance curve with evaluation of areas and horizontal baseline

ever, near the solvent front in TLC there is a positive slope in the baseline and integrators will find a positive systematic error. Figs. 8 and 9 give an illustration of this error exemplified by the marked area F_4. Assuming a horizontal baseline (Fig. 9) the area would be about 2% larger.

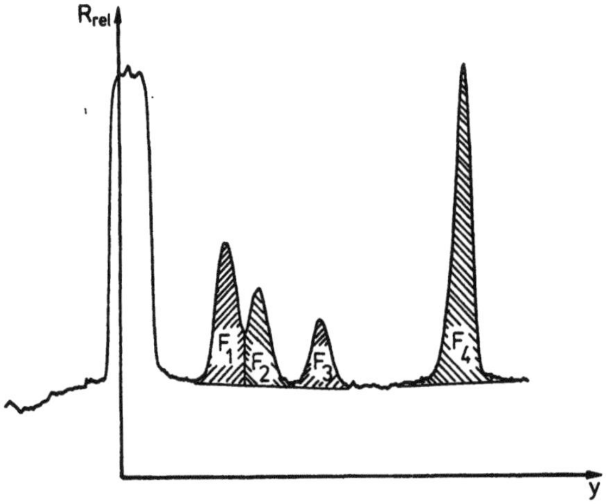

Fig. 9. Recorded reflectance curve with evaluation of areas using sloped baselines

In the case of overlapping peaks systematic errors in integration procedures will similarly arise. If the end of integration of the first peak (marked F_1 in Fig. 9) and the start of integration of the second peak (marked F_2 in Fig. 9) is taken as the minimum of the reflectance curve between the two peaks there is nearly no error if both peaks have the same magnitude. However, the systematic error will increase with an increasing difference between the two areas; the value of smaller area will be too high. In such cases another integration technique may be used. Instead

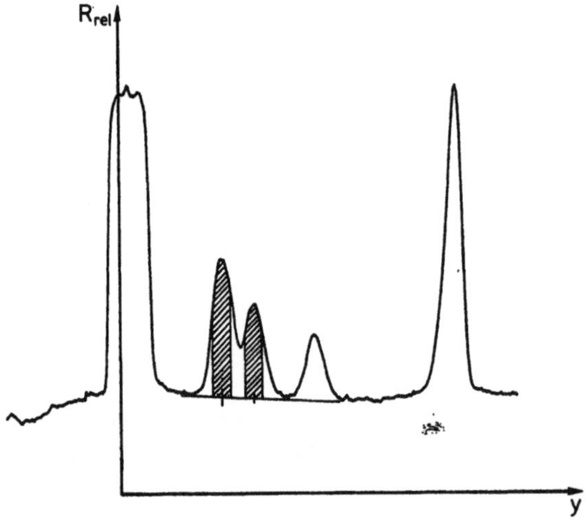

Fig. 10. Recorded reflectance curve with evaluation of fractionalized areas for determination of overlapping peaks [19]

of integrating the whole area, only a "sliced area" will be determined as shown in Fig. 10. This technique has been used with good results in quantitative infrared [20,21] ultraviolet spectroscopy [22] and in multicomponent analysis. Sensitivity can be inreased by using HPLC in connection with photodiode array detectors [23]. In TLC computer controlled instruments, as introduced by our group [24,25,26] have the same effect. Comparing Fig. 10 with Fig. 7 shows that fractionalized areas will only seldomly be influenced by the neighbouring peaks.

2.3 Spline Interpolation

Recorded reflectance curves contain the entire information of the space dependent signal measuring a TLC-plate. Digitizing will lead to an increasing loss of information with decreasing number of data points. Instead of connecting the digitized data by straight lines to a polygon as shown in Fig. 11 C, splines can be used to interpolate between these supporting points by means of polynominals of third degree (Fig. 11 D). Splines have the significant advantage of minimizing the changes of slope [28]. The first hint for applying splines to TLC data was given by Kaiser [29].

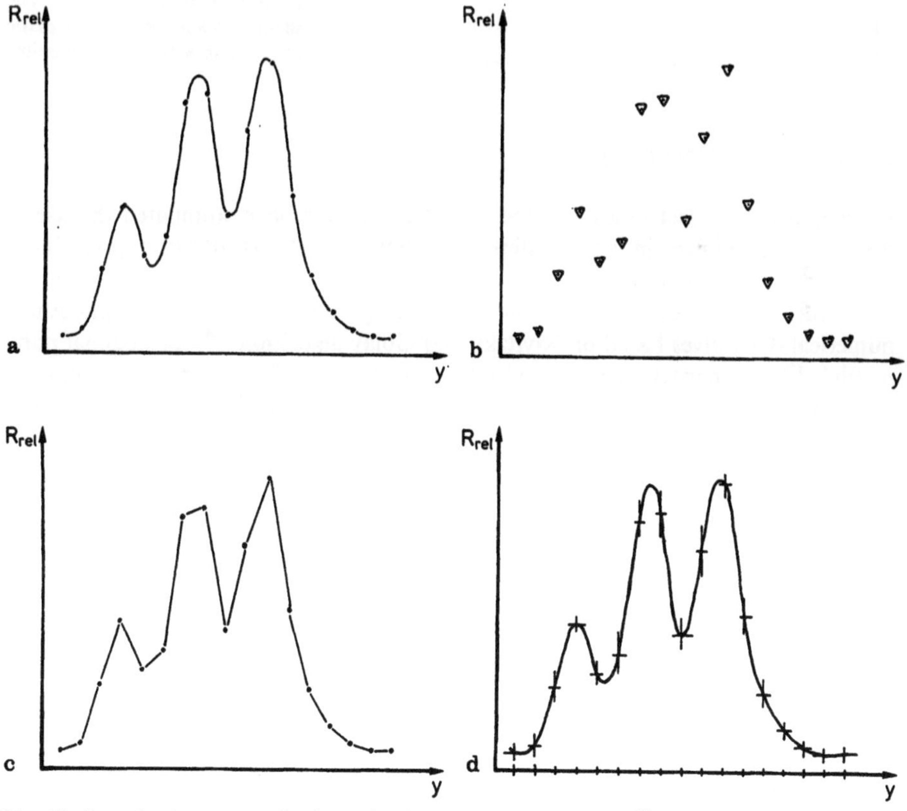

Fig. 11. Introduction to application of splines to chromatograms [30]. **a** measured curve with digitized data; **b** digitized data only; **c** polygonal interpolation; **d** spline interpolation

Polygonal approximations only make use of the coordinates x_i and y_i of each measured data, while splines employ the fact that a measured curve has the same slope and the same change of slope at the supporting point of two adjacent intervals. Spline interpolations may be used for integration purposes but cannot be used for determining peakheights. Moreover, splines can be used for constructing curved baselines in order to avoid systematic errors in integration of peaks near the solvent front as shown in Fig. 12.

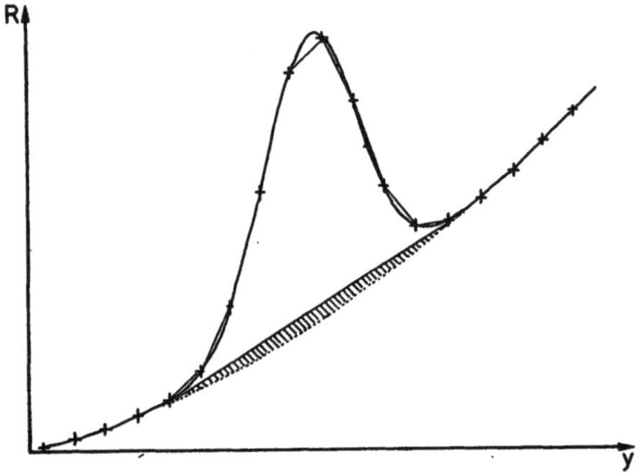

Fig. 12. Baseline approximation by interpolating splines [30]. The shaded area marks the systematic error using a tangent as baseline

2.4 Derivative Recording

Developments in electronics and the use of computers have stimulated the use of derivative techniques in such different analytical fields as titration procedures, polarography (differential pulse polarography) or spectroscopy. In TLC and HPTLC first order derivatives were introduced by our group [26, 31, 32, 33] using smoothing numerical derivatives based on Savitzky and Golay algorithms [34] (orthogonal polynomials [35]) in connection with computer-controlled TLC-scanners. Gelpi and coworkers used electronic circuits for derivative recording in TLC up to fourth order derivatives [36, 37, 38].

In quantitative analysis derivative recording yields an improved detectability of curve data. Comparison of Fig. 6 with its derivative Fig. 13 shows that peak 5 contains another substance in addition to chrysophanol. Derivatives cannot increase the content of information but they may help to identify the total available information especially in overlapping peaks.

In quantitative analysis derivative recording can be used as well. The peak width is determined by the chromatographic process and, thus, the slope depends on the amount of substance per spot. In case of a minor overlap, the same conditions of calibration as in normal TLC/HPTLC are to be applied (cf. part 3); strongly overlapping peaks require non-linear calibration even with low amounts of substance [33]. A further advantage of derivative recording in quantitative TLC is the suppression of baseline effects, a problem expecially arising in TLC of drugs in body fluids or in the analysis of naturally occurring drugs.

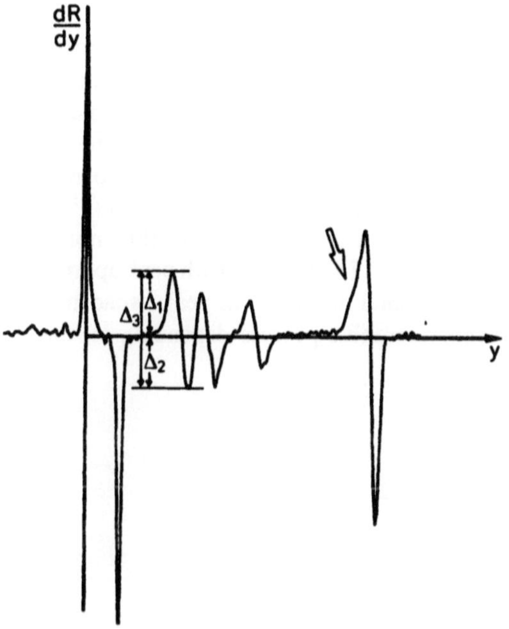

Fig. 13. First order derivative recording of the same chromatogram as shown in Fig. 6 [19]

One complication of derivative recording should be mentioned here in order to avoid errors in quantitative analysis. Each derivative technique, by an electronic or numeric procedure, is a source of systematic errors. In the case of electronically generated derivatives, the error function has the shape of the next higher order derivative; numerically generated derivatives show a different shape. It is important to notice, however, that the systematic error increases in a geometric series magnitude 1:2:4:8:16 [39]. Moreover, it has to be taken in account that the noise in derivatives increases with the faculty series 1:2:6:24:120 [40]. To diminish the noise in electronically generated derivatives damping circuits are used, while in numerically generated derivatives the approximation range is extended. Both procedures will lead to larger systematic errors. Therefore, in precision analysis only first or second order derivatives should be used.

2.5 Peak Approximation

Peak approximation techniques have been used in GC for more than 25 years in order to resolve fused peaks [41] and other curve anomalies of chromatograms. In most cases, gaussian profiles of the peaks are proposed and a nonlinear curve fitting procedure is applied to fit the data and to integrate the peaks, skew peaks included [42,43]. All procedures are actually adapted, i.e. they do not take in account the typical band broadening with time (subsequent peaks will increase in peak width). Therefore, malfunctions of the published programs are possible, as shown in the paper of Littlewood [44]. Most papers, even those published further back make notice of peak distortion [45,46]. Nonlinear regression techniques are used to fit the raw data to the model function.

In contrast to GC, there are very few investigations in this field of TLC. One of the reason may be that many investigators had not been successful in transferring the mathematics of GC to TLC. Another problem arises from the instrument: either the noise is to high or the peak is distorted by the time constant of the smoothing RC-circuit. Using a computer controlled TLC-scanner [24,25,26] (cf. I. Böhrer in this volume) it is possible to measure a nonmoving plate so that peak approximation procedures can successfully be employed to evaluate nonresolved peaks [31,47]. A typical result is given in Fig. 14. In this case, a special gaussoid (but not really gaussian) model function had been used. To date, TLC research with peak approximation has not made use of TLC theory (binominal distribution, peak broadening by spotting, peakwidth caused by the chromatographic process depending on the R_f-value, peak-broadening by diffusion).

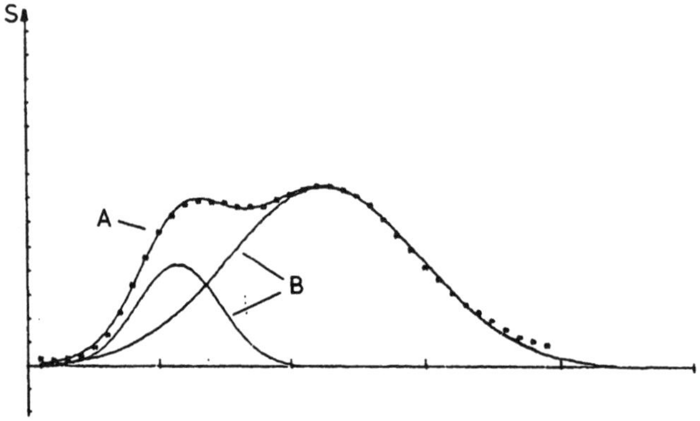

Fig. 14. Peak approximation of two unresolved peaks in TLC. Substances: Glucofrangulin A and B

3 Calibration

Reflectance or fluorescence measurements in TLC signals, whether peakheight or integrated area, are strongly influenced by various experimental details. Layer thickness, mean particle size, distribution of particle size, migration distance and R_f-values will change the reflectance curve, although under different experimental conditions, such as spotting or setting, the parameters of the scanner are held constant. Therefore, it is necessary to carry out calibration measurements on each plate. All types of regression techniques have been applied to TLC data because of the nonlinear relationship between signal and amount of substance per spot in a larger working range.

3.1 Linear Regression

As discussed briefly in section 2.1 nearly linear relationship should be used for calibration only at low amounts of substance per spot ($m < 0.3$ µg/spot) [12,13,14,15], independent of the signal used, such as transmittance, reflectance or logarithms of these [8,15]. By using double beam ratio recording instruments — the signal following log R_x/R_0 — the linear range will in some cases be extended to higher concentrations, occassionally up to 1–2 μg/spot [48,49]. In routinized quality control of industrial or pharmaceutical products it is also possible to attain an extended working range, for instance, from 80% to 120% of the expected amount, with a nearly linear relationship [50,51]. In these cases a straight line (14) is used for calibration. To use all possible information from the measurement the straight line must be calculated by means of the linear regression technique. However, there are some restrictions we must observe: the errors of the data lie within a normal distribution and the variations of the data must be constant within the calibration range. Linear regression is further based on the theorem that the independent variable m is without error.

$$R = a_0 + a_1 m \tag{14}$$

R reflectance

m amount of substance per spot

In TLC errors of measured data seem to be within a normal distribution [52] based on the test of David [53,54a]. Furthermore, in wide range calibration with, e.g. papaverine m = 2 ng/spot up to m = 2 μg/spot, variances are constant if measurements are made on the same plate [52] using the test designed by Cochran [55,54b]. If the errors do not find themselves within the normal distribution, linear regression is possible, but error propagation will lead to smaller predictable errors than obtained in the actual results. In the absence of constant variance the linear regression can fail and give significantly, wrong results. In TLC, the theorem of independent variables without errors is not valid as of the error of spotting *small volumes*. Instead of linear regression a linear correlation may be made [56]. Experience remarkably shows, however, that there is nearly no difference between both methods.

All uncertainties of calibration and errors of the measurement lead to an incorrect analytical result. It is possible to calculate the confidence interval of the calibration

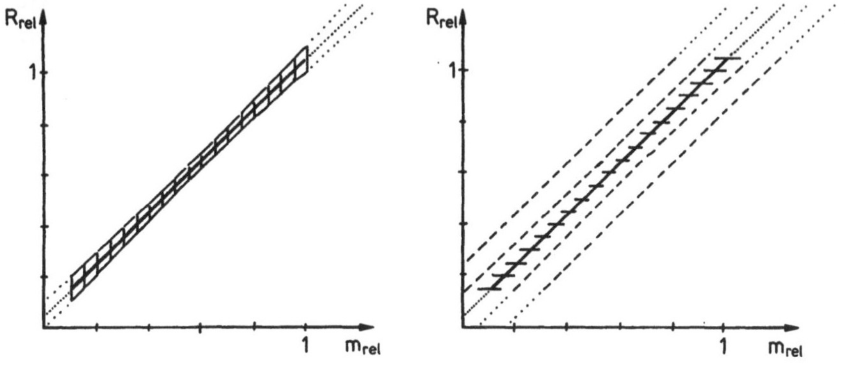

Fig. 15. Straight line calibration with confidence interval of calibration defined in y direction **a** and confidence intervals of the analytical result defined in x-direction **b** applying variing amounts of repeated measurements in the analytical procedure [58]

which is usually defined in direction of the dependent variable [57] as shown in Fig. 15 A. Of more concern to the analyst is the confidence interval of the analytical result defined in the direction of the independent variable (Fig. 15 B) [58]. In the latter case, the number of independent measurements of the analysis also define the confidence interval.

3.2 Regression with Linearization

Normally, in wide range calibration there is a nonlinear relationship between the measured signal and the amount of substance per spot. The various publications describe different linearization procedures. One of the most frequently used methods square the measured peakheight or area under the curve, as proposed by Tausch [11]. Some authors take the logarithms of R [59], the reciprocals of R [60,61,62] or the reciprocals of R and m [63,64] as the base for linear regression. Kaiser [63] emphasized the linear regression with log R vs. log m for linearization of reflectance data, as introduced by Otteneder and Hezel [65] in a non-linear relationship in fluorescence measurements in case of reabsorption. Moreover, a parabolic regression after forming the logarithms of both variables has been published [66] and emphasized by Schmutz [67].

The problem in all these linearization procedures is that the most important condition for any type of regression is not fulfilled: the steadiness of variances. Each transformation of R-data will lead to differing variances, the change of variances being directly dependent on the transformation function. While for log R the transformed variance decreases, the transformed variance of the reciprocal 1/R increases with increasing m.

On the other hand, a transformation of the independent variable does not change the variance. Therefore, transformations of m into log m, as described by Pataki [69] for fluorescence or by Connors [70], Huber [12] and Koleva [71] for reflectance are possible bases for linear regression analysis. Instead of calibration with (15) the inverse transformation (16) should be preferred towards a correct regression.

$$R^2 = k \cdot m \tag{15}$$

$$R = k \cdot \sqrt{m} \tag{16}$$

A multivariate linear regression can successfully be employed for moderately wide range calibration. As an example parabolic regression (17) [31,72] should be mentioned. Higher order polynomials have disadvantages in terms of degree of freedom and error propagation, and the results are less precise.

$$R = a_0 + a_1 \cdot m + a_2 \cdot m^2 \tag{17}$$

3.3 Non-linear Regression

In 1913 Michaelis and Menten derived an equation describing the non-linear behaviour of the reaction rate of a substrate-specific enzyme. At low concentrations the velocity of the reaction is nearly linear, while saturation occurs at higher concentrations. For further details textbooks of biochemistry may be consulted. The same conditions are valid in calibration of TLC data: at low amounts of substance per spot there is a linear relationship at higher amounts the sensitivity decreases. In terms of reflectance measurement, the Michaelis-Menten equation is written in the form (18):

$$R = \frac{R_{max} \cdot m}{K + m} \tag{18}$$

R_{max} maximum reflectance
K constant describing the non-linearity

In former days, before electronic caclulations came into existance all calibrations and evaluations had to be carried out manually by graphical methods; Linearized solutions had been used instead of nonlinear regression. Lineweaver and Burk [73] derived the linearized equation (19). It was introduced for calibration purposes to TLC by Kufner and Schlegel [74]. Kaiser [63], Hulpke and Stegh [64] also used calibration techniques with reciprocal transformations of R and m without reference to the Michaelis-Menten transformation.

$$z = b_0 + b_1 \cdot u \tag{19}$$

$$z = \frac{1}{R} \tag{20}$$

$$u = \frac{1}{m} \tag{21}$$

$$b_0 = \frac{1}{R_{max}} \tag{22}$$

$$b_1 = \frac{K}{R_{max}} \tag{23}$$

Another type of linearization was introduced by Eadie [75]. In this case the dependent variable R is not transformed, whereas the independent variable m is replaced by R/m (25). This procedure has also been discussed by Kufner and Schlegel and is routinely used by Sistovaris [76, 77, 78]:

$$R = R_{max} - K \cdot u \tag{24}$$

$$u = \frac{R}{m} \tag{25}$$

The problem in both methods is the error propagation. If an error exists in the measurement, this error will be submitted to the transformation as well. A second problem arises in the variances. Usually the variances of measurement in TLC are constant within the calibration range. The transformation of data will lead to inhomogeneous variances and this is the reason for unreliable regression analysis.

The Lineweaver-Burk-transformation on papaverine for instance yields the calibration shown in Fig. 16. The scattering of the original data is distributed normally, according to the test designed by David [53, 54a)] and homogeneous variances exist, according to the test developed by Cochran [55, 54b)]; however, it is easy to see form Figs. 16 and 17 that this is not the case after transformation.

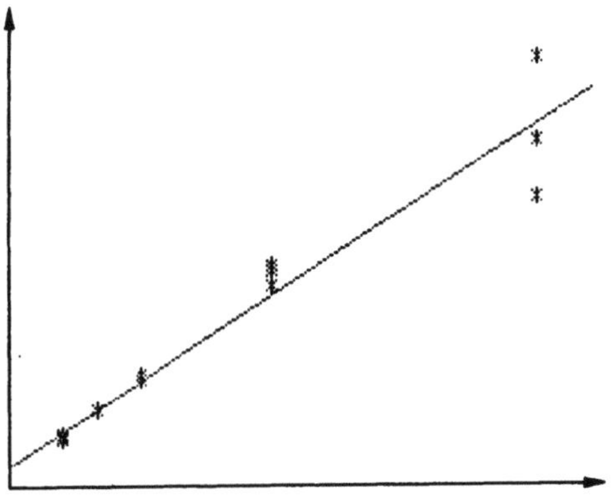

Fig. 16. Michaelis-Menten-function: Linearized calibration according to Lineweaver-Burk [73)] in TLC [79)]. Substance: papaverine HCl; working range: 4–40 ng/spot

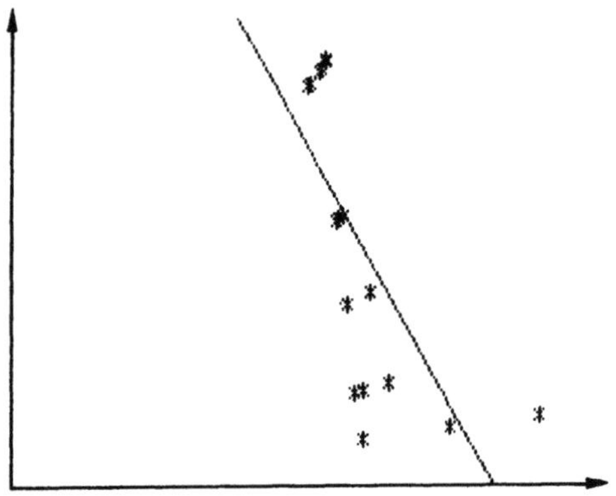

Fig. 17. Michaelis-Menten-function: Linearized calibration according to Eadie [75)] in TLC (same plate as in Fig. 16) [79)]

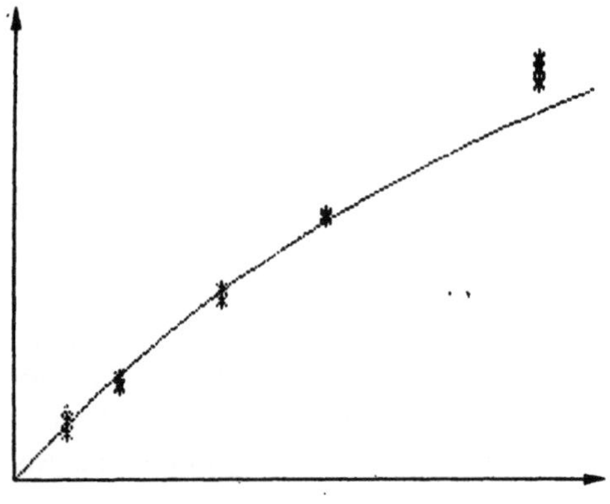

Fig. 18. Michaelis-Menten-function for use in calibration in TLC: Recalculated calibration curve from linearized regression according to Lineweaver-Burk [73)] (cf. Fig. 16) [79)]

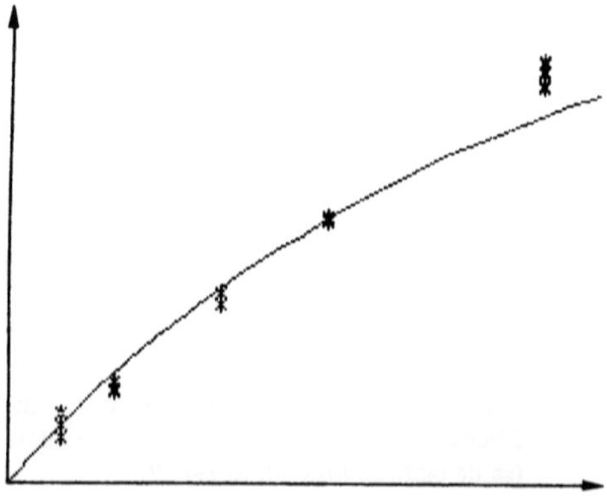

Fig. 19. Michaelis-Menten-function for use in calibration in TLC: Recalculated calibration curve from linearized regression according to Eadie [75] (cf. Fig. 17) [79]

The wrong results of these linearized regressions are evident from Fig. 18 and Fig. 19. From the results of the linearized regression data R_{max} and K had been calculated. The calibration curve was plotted using these coefficients [79].

More reliable results are obtained with a non-linear regression of the data by directly using Eq. (18) [79]. Without referring to the mathematrical background an estimate of the coefficients R_{max} and K for instance by the Michaelis-Menten-Eadie-method, can be used to calculate the refinements δR_{max} and δK using matrix operations and the newtonian matrix (29). From this the refined coefficients are calculated (30), (31), and with an iterative procedure R_{max} and K are determined up to a preset range of precision.

$$A_i = \frac{\partial R_i}{\partial R_{max}} = \frac{m_i}{K + m_i} \tag{26}$$

$$B_i = \frac{\partial R_i}{\partial K} = -\frac{R_{max} \cdot m_i}{K + m_i} \tag{27}$$

$$R_i = \frac{R_{max} \cdot m_i}{K + m_i} \tag{28}$$

$$\begin{pmatrix} \sum A_i^2 & \sum A_iB_i \\ \sum A_iB_i & \sum B_i^2 \end{pmatrix} \begin{pmatrix} \delta R_{max} \\ \delta K \end{pmatrix} = \begin{pmatrix} \sum R_iA_i \\ \sum R_iB_i \end{pmatrix} \tag{29}$$

$$R_{max} = R_{max} + \delta R_{max} \tag{30}$$

$$K = K + \delta K \tag{31}$$

The result of the non-linear regression is shown in Fig. 20. Comparing with the result of the linearized regression, it is obvious that there is no systematic error and that the results are more precise.

Table 1 allows a comparison of the results of R_{max} and K. The advantage of the nonlinear regression can be seen using the approximate standard deviation of R (32)

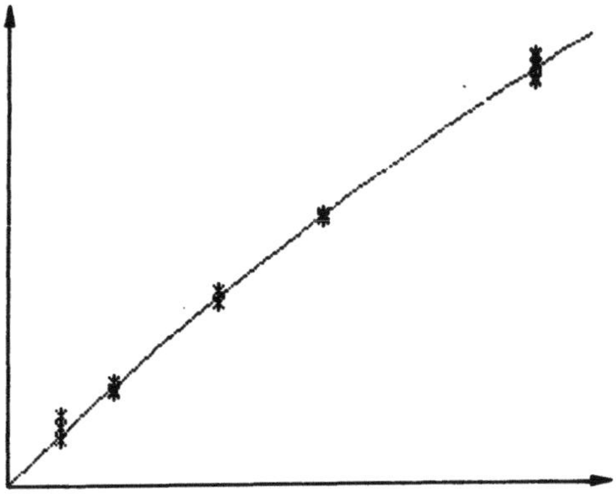

Fig. 20. Non-linear regression to fit the Michaelis-Menten-function to TLC-data (same plate as Figs. 16 to 19) [79]

or the favourable fit (33), proposed by Stork and Knecht [80] and Kaiser [63].

$$\mathrm{var}\,(R) = \sqrt{\frac{\sum (R_i - \hat{R}_i)^2}{n-2}} \tag{32}$$

$$G_K = \frac{1}{n} \sum \frac{|R_i - \hat{R}_i|}{R_i} \tag{33}$$

Table 1. Results of linearized and nonlinear regression. Substance: papaverine; precoated HPTLC-plate, E. Merck, Darmstadt. Solvent: Acetone/toluene 20:20 (v/v); Camag TLC-scanner, reflectance at 254 nm [79]

	Linear regression		nonlinear regression
	Lineweaver-Burk	Eadie	
R_{max}	512.8	455.2	1039.1
K	66.1	54.0	153.3
var (R)	3.2	3.2	1.1
G_K	0.0821	0.0965	0.0498

4 Standard Techniques

Instead of calibration chromatographic analysis often involves standard techniques as the external or internal standard method. Both methods use spotting as shown in Fig. 21. For the simplest case of TLC, calibration and analysis are alternating and each analysis is related to the preceding track (21 A) or to the mean of the preceding and succeeding track (21 B). If gradients exist in layer thickness or particle size distribution the data pair technique according to Frei [81] should be used to diminish systematic errors. In this case, double spotting, as shown in Fig. 21 C, is preferably accomplished and the mean is taken from the two. In HPTLC layer thickness and particle size distribution are more homogeneous and, therefore, the number of standards may be decreased; a mean of all standards (Fig. 21 D) is taken for evaluation. Standard addition techniques have no importance in TLC/HPTLC much in contrast to other analytical methods, such as atomic absorption spectroscopy.

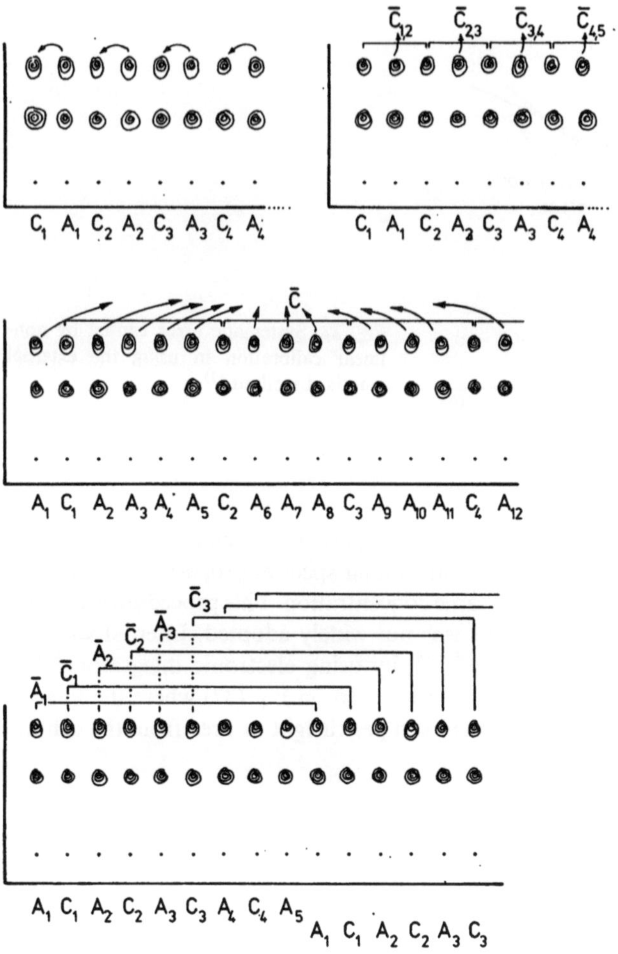

Fig. 21. Spotting in TLC/HPTLC using the external and internal standard method (explanation in the text)

4.1 External Standard Method

The external standard method is the most frequently used type of evaluation in routine TLC. Evaluations are carried out by the simple Eq. (34). The basic conditions of this evaluation technique are firstly, a linear relationship between the signal peakheight or area and the amount of substance per spot and secondly, the fact that no blanks are necessary. A straight calibration line leads through the origin. If nonlinear calibration curves are used the analytical result will contain a systematic error as demonstrated in Fig. 22.

$$m_{anal} = m_{cal} \cdot \frac{R_{anal}}{R_{cal}} \tag{34}$$

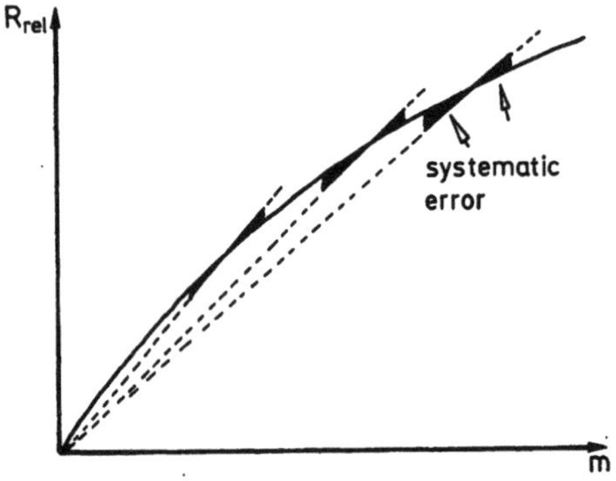

Fig. 22. Systematic error caused by nonlinear calibration in using the external standard method [47)]

4.2 Internal Standard Method

The internal standard method is based on the comparison of signals of the substance to be determined and another substance with similar spectral properties mixed to the calibration and analysis solution in the same concentration. This procedure was introduced to TLC by Klaus in 1972 [82)] but was not widely adopted. Internal standard techniques lead to more precise results [87,88)] by using electronic integrators [83)] or automated systems [84,85,86)]. The calculation is based on Eq. (37) where the internal standard factor is calculated from the results in peakheight or area from the calibration track according to Eq. (36).

$$m_{x,\,anal} = m_{s,\,anal} \cdot \frac{R_{x,\,anal}}{R_{s,\,anal}} \cdot f \tag{35}$$

$$f = \frac{m_{x,\,cal}}{m_{s,\,cal}} \cdot \frac{R_{s,\,cal}}{R_{x,\,cal}} \tag{36}$$

The internal standard technique is based on the covariance term in the error propagation of this method [89)] and avoids the error caused by applying small volumes to the plate. But there is another important aspect of error propagation in the comparison of internal and external standard methods. If the error in measurement is smaller than the error in spotting, the internal standard method is more precise than the external standard method [90)]. However, if the error in measurement becomes dominant, the external standard leads to better results. This is the reason why in early quantitative TLC, where peaks had been triangulated and evaluated by the squared area with all problems of an infavourable error propagation [91], the internal standard technique could not be adopted. For this method straight calibration lines leading through the origin are also a prerequisite. A further disavantage is that one additional substance must be separated in the chromatographic process.

4.3 Standard Addition Method

The standard addition method is commonly used in quantitative analysis with ion-sensitive electrodes and in atomic absorption spectroscopy. In TLC this method was used by Klaus [92]. Linear calibration with $R(m=o)=o$ must also apply for this method. However, there is no advantage compared with the external standard method; even worse there is a loss in precision by error propagation. The attainable precision is not satisfactory and only in the order of 3–5%, compared to 0.3–0.5% using the internal standard method [93].

5 Limit of Detection and Determination

The terms limit of detection and limit of determination have been confusingly used in the literature. While limit of detection is a qualitative value, the term limit of determination should only be used in connection with quantitative work. The detection of an impurity is essential in routine analysis but it is quite different from determining the content of the impurity.

5.1 Limit of Detection

The limit of detection is the smallest amount of a substance detectable by an analytical method. This limit is strongly related to the noise of the measuring process. The definition given by Kaiser [94,95] in the field of spectroscopy [96] should be understood in these terms. The most frequently used definition is stated by Eq. (37) where sdv(b) is the estimated standard deviation of a blind, i.e. the noise of the analytical procedure:

$$m_{dtc} = 2 \cdot \mathrm{sdv}(b) \qquad (37)$$

In TLC the noise of measurement is smaller than the irregularities of the plate or texture of the sorbens layer caused by microscopic changes in the particle size distribution. Therefore, the limit of detection can be lowered by scanning the plate before development by a computer-coupled scanner, storing this background and correcting the real scan by these data [97]. The limit of dection depends on most parameters of TLC, such as volume to be spotted, migration distance, R_f-value, spectral data of the substance, measuring technique employed and others [15].

5.2 Limit of Determination

Each quantitative determination should be seen in context with the probability of error (e.g. $\alpha = 0.05$) and the confidence interval of the result (e.g. $p = 0.95$). Therefore, the limit of determination depends not only on the noise of measurement but also on the entire calibration procedure. In consideration of this, the limit of determination is defined as the smallest amount of a substance to be quantitativly determined and distinguished from the amount m=o or c=o by statistical probability (e.g. p = 0.95). Consequently, also the measured signal has to be known with the proposed statistical probability, meaning the limit of detection is only to be specified by repeated measurements in connection with the real statistics of the calibration curve [98]. The statistic definition of the limit of determination is explainable by Fig. 23 using the confidence interval in the analytical result (cf. Fig. 15) [98].

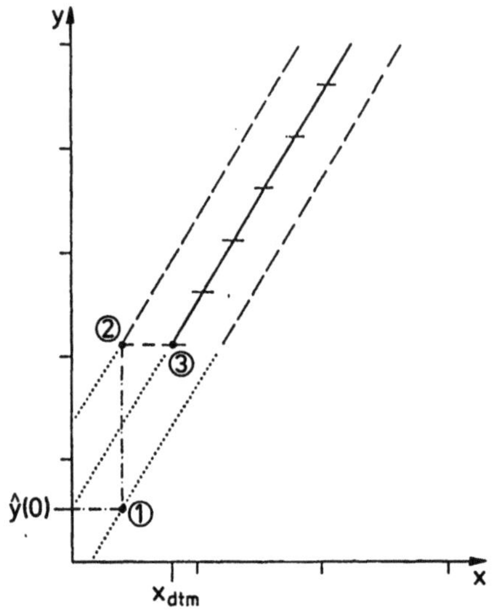

Fig. 23. Statistical definition of the limit of determination by the confidence interval in the analytical result [98]

The limit of determination depends on the precision of the measurement sdv(R), on the sensitivity of the measurement (the slope of the calibration curve), on the number of data used for calibration n_c and analysis n_a, on the choice of calibration points (m_{cal}) and on the statistical probability influencing students t-factor. Proceeding from $n_c = 4$ calibration data on a straight line to $n_c = 8$ the limit of determination is lowered by a factor of four. In analysis a threefold determination is sufficient [98].

6 Acknowledgements

The author kindly acknowledged the generous support of his research by the Deutsche Forschungsgemeinschaft, which was partially reviewed in this paper. Financial support given by the Fonds der Chemischen Industrie is greatly appreciated. Furthermore, the author wishes to express his gratitude to all his coworkers in the field of TLC for their engagement and their activities.

7 References

1. Kortüm, G.: Reflexionsspektroskopie, Springer Verlag, Berlin u. a. O. 1969
2. Chandrasekhar, S.: Radiative Transfer, Calderon Press, Oxford 1950
3. Schuster, A.: Astrophys. J. *21*, 5 (1905)
4. Ebel, S., Post, P.: J. High Res. Chrom. *4*, 337 (1981)
5. Kubelka, P., Munk, F.: Z. Techn. Phys. *12*, 539 (1931)
6. Kubelka, P.: J. Opt. Soc. Amer. *38*, 448 (1948)
7. Bodó, Z.: Acta Phys. Acad. Sci. Hung. *1*, 135 (1951)
8. Post, P.: Ph. D. Thesis, Marburg 1979
9. Prosek, M. et al.: J. High Res. Chrom. *2*, 517, 661 (1979)
10. Prošek, M. et al.: ibid. *3*, 183 (1980)
11. Tausch, W.: Meßtechnik *80*, 38 (1971)
12. Huber, W.: J. Chromatogr. *33*, 378 (1968)
13. Bethke, H., Frei, R. W.: ibid. *91*, 433 (1974)
14. Kußmaul, H.: Ph. D. Thesis, Marburg 1972
15. Ebel, S., Herold, G., Hocke, J.: Instrum. Forschg. *1975*, 42
16. Brenner, M., Niederwieser, A., Pataki, G., Weber, R.: Dünnschichtchromatographie (ed. Stahl, E.) Springer Verlag, Berlin u. a. O. 1962[1] (1972[2] does not include this part)
17. Ebel, S., Klarner, D.: to be published
18. Ebel, S., Geitz, E., Klarner, D.: Kontakte (Merck) *1980* (1), 11
19. Alert, D., Ebel, S.: unpublished results on analysis of anthraquinone containing drugs
20. Ebel, S., Walter, V., Weidemann, E.: paper presented at Pittsburgh Conf. 1981
21. Weidemann, E.: Ph. D. Thesis, Marburg 1983
22. Ebel, S., Bender, R., Weyandt, M., Widjaja, N.: Dtsch. Apothek.Ztg. 123, 2151 (1983)
23. Ebel, S., Werner-Busse, A.: paper presented at TLC/HPLC meeting, Saarbrücken 1983 (cf. Fresenius Z. Anal. Chem. in press)
24. Ebel, S., Hocke, J.: J. Chromatogr. *126*, 449 (1976)
25. Ebel, S., Hocke, J.: Chromatographia *10*, 123 (1976)
26. Ebel, S., Geitz, E., Hocke, J.: p. 55 in [27)]
27. Bertsch, W. et al. (Edit): Instrumentalized HPTLC, Hüthig Verlag, Heidelberg 1980
28. Späth, H.: Spline-Algorithmen zur Konstruktion glatter Kurven und Flächen, Oldenbourg Verlag, München 1973
29. Kaiser, R. E.: paper presented at TLC Symp. Darmstadt 1981
30. Alert, D., Ebel, S., Schaefer, U.: unpublished results
31. Ebel, S., Geitz, E., Hocke, J.: GIT Fachz. Laborat. *24*, 660 (1980)
32. Geitz, E.: Ph. D. Thesis, Marburg 1981
33. Alert, D., Ebel, S., Geitz, E., Schaefer, U.: J. Pharm. Biomed. Anal in press
34. Savitzky, A., Golay, M. J. E.: Anal. Chem. *36*, 1627 (1964); corrections cf.: Steinier, J., Termonia, Y., Deltour, J.: Anal. Chem. *44*, 1906 (1972)
35. Ludwig, R.: Methoden der Fehler- und Ausgleichsrechnung; Vieweg Verlag, Braunschweig 1969
36. Traveset, J., Such, V., Gonzalo, R., Gelpi, E.: J. Chromatogr. *204*, 51 (1981)
37. Such, V., Traveset, J., Gonzalo, R., Gelpi, E.: ibid. *234*, 77 (1982)
38. Traveset, J., Such, V., Gonzalo, R., Gelpi, E.: p. 233 in [68)]
39. Ebel, S.: Proc. of. 1. Gesamtkongress Pharmazeut. Wissenschaften, München 1983
40. Walter, V.: Ph. D. Thesis, Marburg 1983
41. Fraser, R. D. B., Suzuki, E.: Anal. Chem. *38*, 1770 (1966)
42. Gladney, H. M., Dowden, B. F., Swalen, J. D.: ibid. *41*, 883 (1969)
43. Roberts, S. M., Wilkinson, D. H., Walter, L. R.: ibid. *42*, 886 (1970)
44. Anderson, A. H., Gibb, T. C., Littlewood, A. B.: ibid. *42*, 434 (1970)
45. Grushka, E. et al.: ibid. *41*, 889 (1969); *42*, 21 (1970)
46. McWilliam, I. G., Bolton, H. C.: ibid. *41*, 1755 (1969)
47. Kaal, M.: Ph. D. Thesis, Marburg, 1979
48. Penner, M.: J. Pharm. Sci. *57*, 2132 (1968)
49. Knappstein, P., Touchstone, J.: J. Chromatogr. *37*, 83 (1968)
50. MacMullen, E. A., Heveran, J. E.: in Quantitative Thin-Layer-Chromatography, (Touchstone, J. C. Ed.), Wiley, New York 1973

51. Ebel, S., Herold, G.: Fresenius Z. Anal. Chem. *273*, 7 (1975)
52. Alert, D., Ebel, S., Jork, H., Schaefer, U.: unpublished results
53. David, H. A., Hartley, H. D., Pearson, E. S.: Biometrika *41*, 482 (1954)
54. Sachs, L.: Angewandte Statistik, Springer Verlag, Berlin 1978; 54a: p. 253; 54b: p. 383
55. Cochran, W. G.: Ann. Eugen. *11*, 47 (1941)
56. Ebel, S., Geitz, E., Glaser, E.: J. High Res Chrom. *4*, 508 (1981)
57. Doerffel, K., Eckschlager, K.: Optimale Strategien in der Analytik, Deutsch Verlag, Thun, Frankfurt 1981
58. Ebel, S.: Comput. Anwend. Laborat. *1*, 56 (1983)
59. Frodyma, M. M., Frei, R. W., Williams, D. J.: J. Chromatogr. *13*, 61 (1964)
60. Klaus, R.: ibid. *16*, 311 (1964)
61. Stahl, E., Jork, H.: Zeiss Informat. *16*, 52 (1968)
62. Röder, K., Eich, E., Mutschler, E.: Arch. Pharmaz. *304*, 297 (1971)
63. Kaiser, R. E.: in [27] p. 165
64. Hulpke, H., Stegh, R.: in [27] p. 113
65. Otteneder, H., Hezel, U.: J. Chromatogr. *109*, 181 (1975)
66. Müller, M.: Chromatographia *13*, 557 (1980)
67. Schmutz, H. R.: in [68] p. 246
68. Kaiser, R. E. (Edit.): Instrumental HPTLC Proc. Interlaken 1982, Inst. Chromatographie, Bad Dürkheim 1982
69. Pataki, G., Kunz, A.: J. Chromatogr. *23*, 465 (1966)
70. Connors, W. M., Boak, W. K.: ibid. *16*, 243 (1964)
71. Koleva, M., Joneidi, M., Budenski, O.: Pharmazie *28*, 199, 317 (1973)
72. Ebel, S., Herold, G.: Dtsch. Lebensm. Rundsch. *70*, 133 (1974)
73. Lineweaver, H., Burk, D.: J. Amer. Chem. Soc. *56*, 658 (1934)
74. Kufner, G., Schlegel, H.: J. Chromatogr. *169*, 141 (1979)
75. Eadie, G. S.: J. Biol. Chem. *146*, 85 (1942)
76. Sistovaris, N.: GIT Fachz. Laborat. Suppl. Chromatogr. *1983* (3), 17
77. Kark, B., Sistovaris, N., Keller, A.: J. Chromatogr. 275, 188 (1983)
78. Sistovaris, N.: ibid. 276, 139 (1983)
79. Alert, D., Ebel, S., Schaefer, U.: Chromatographia *18*, 23 (1984)
80. Knecht, J., Stork, G.: Fresenius Z. Anal. Chem. *270*, 97 (1974)
81. Bethke, H., Santi, W., Frei, R. W.: J. Chromat. Sci. *12*, 392 (1974)
82. Klaus, R.: J. Chromatogr. *62*, 99 (1972)
83. Ebel, S., Herold, G.: Fresenius Z. Anal. Chem. *270*, 19 (1974)
84. Ebel, S., Herold, G., Hocke, J.: Chromatographia *8*, 573 (1975)
85. Ebel, S., Hocke, J.: ibid. *9*, 78 (1976)
86. Pohl, U., Schweden, W., Lessnig, W., Metz, G.: Fresenius Z. Anal. Chem. *285*, 111 (1977)
87. Ebel, S., Herold, G.: Chromatographia *8*, 35 (1975)
88. Ebel, S., Herold, G.: ibid. *8*, 569 (1975)
89. Haefelfinger, P.: Fresenius Z. Anal. Chem. 218, 73 (1981)
90. Ebel, S., Glaser, E., Rost, D.: J. High Res. Chrom. *2*, 250 (1978)
91. Ebel, S., Hocke, J.: ibid. *2*, 156 (1978)
92. Klaus, R.: J. Chromatogr. *40*, 235 (1965)
93. Ebel, S., Geitz, E.: Kontakte (Merck) *1981* (2), 34
94. Kaiser, H.: Fresenius Z. Anal. Chem. *209*, 1 (1965)
95. Kaiser, H.: ibid. *216*, 80 (1966)
96. Kaiser, H.: Spectrochim. Acta *3*, 40 (1947)
97. Kaiser, R. E., Rieder, R. I., Miller, A., Pilgram, B.: in [68] p. 335
98. Ebel, S., Kamm, U.: Fresenius Z. Anal. Chem. *316*, 382 (1983)

Evaluation Systems in Quantitative Thin-Layer Chromatography

Irmgard M. Böhrer

c/o L. Heumann & Co. GmbH, Heideloffstraße 18–28, 8500 Nürnberg, FRG

Table of Contents

1 Introduction 97

2 Construction Principle and Equipment of the Instruments 97
2.1 Construction Principle 97
2.2 Equipment of the Instruments 98
2.2.1 Reflectance Mode of Scanning 98
2.2.2 Transmission Mode of Scanning 99
2.2.3 Simultaneous Reflectance and Transmission Mode of Scanning . . 99
2.2.4 Fluorescence Mode of Scanning 100

3 Evaluation Systems with Different Kinds of Scanning 100
3.1 Slit-beam Densitometers 100
3.2 Point Scanning Densitometers 101
3.2.1 "Flying-spot"-Scanning 102
3.2.2 Zig-zag Scanning 103
3.2.3 Meander Scanning 103

4 Instruments with Different Photometric Systems 103
4.1 Single-beam Densitometers 103
4.2 Double-beam Densitometers 103
4.3 Dual-wavelength Densitometers 104
4.4 Additional Device for Simultaneously Measuring Reflectance and Transmission 104

5 Automatic Systems 105
5.1 Automatic Scanning Instruments 105
5.2 Fully Automatic, Computer-controlled Systems 106
5.2.1 System of the Ebel Group 107
5.2.2 System of the Kaiser Group 110

Topics in Current Chemistry, Vol. 126
Managing Editor: Dr. F. L. Boschke

6 Performance of the Densitometers and Comparison with Instruments of Other Kinds of Chromatography 110
6.1 Advantages and Disadvantages of the "Off-line" System in Thin-layer Chromatography 110
6.2 Sensitivity 112
6.3 Accuracy and Reproducibility 113
6.4 Wear and Maintenance 113
6.5 Time 114
6.6 Personnel and Costs 115
6.7 Fields of Use 115

7 Conclusions 116

8 Acknowledgements 117

9 References 117

1 Introduction

Thin-layer chromatography was not taken seriously as a quantitative method of analysis for a long time. The comparatively simple techniques could in fact be reproduced astonishingly well but were very labour-intensive. It was common practice to scrape the separated spots off the plate together with the adsorbent layer, then to elute and to measure the extracted substances in solution. An improvement on this was the direct elution from the plate. Several elution heads are fastened above the separated spots, solvents are pumped through and the extract is collected in spectrophotometer cells. Thus, several spots can be eluted simultaneously [1].

It was also usual to apply the samples as lines on sheets, to cut out the separated zones and to elute the stripes hung up in an elution appliance by a slow stream of solvents. The method became known as "line elution technique" [2]. Only the possibility, however, of measuring directly on the plate — also called "in situ evaluation" — increased the interest in thin-layer chromatography as a widely used quantitative method. According to Hezel [3] the photometric measuring of thin-layer chromatograms by commercial instruments has been possible since 1967. Since then the instruments, also called densitometers or scanners, have constantly been developed and improved. To evaluate the produced signals one originally depended on registrations of a recorder. The recorded curves were mostly evaluated by measuring the peak height or the peak area (height × half-peak width), now and then they were cut out and weighed; these techniques were time consuming. Integrators had, for the most part, been adapted to gas chromatography and were only gradually made usable for thin-layer chromatography.

Apart from the deficiencies of the measuring devices the available plate material had not been suitable for a long time. With the improvement of the thin-layer plates and, above all, by introducing the high performance thin-layer plates (HPTLC-plates) presented in 1975 by Ripphahn and Halpaap [4], the accuracy of quantitative thin-layer chromatography improved considerably.

Through the automation of the scanning process the method has become more rational and more reliable. New possibilities of evaluation and documentation result from connection to a computer.

In the meantime, thin-layer chromatography has become a very flexible, rational, fast, reliable and precise method of analysis.

2 Construction Principle and Equipment of the Instruments

2.1 Construction Principle

Instruments for the in-situ analysis of thin-layer plates are constructed according to the following principle (Fig. 1). The light from various exchangeable sources passes a monochromator or a monochromatic filter. The monochromatic light falls slit-shaped or spot-shaped onto the thin-layer plate to be measured; the latter is positioned on a movable plate table. The light which is either reflected, transmitted through the plate or emitted in the form of fluorescence finally reaches a detector.

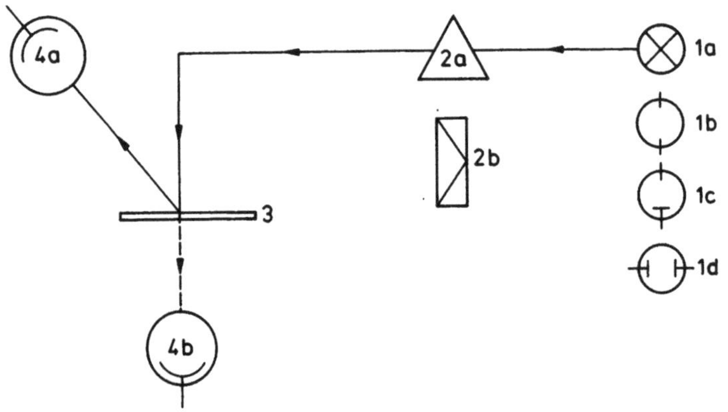

Fig. 1. Construction principle; 1a Tungsten or halogen lamp, b deuterium lamp, c mercury lamp, d xenon lamp; 2a monochromator, b filter; 3 chromatoplate, 4a reflectance detector, b transmission detector

The signal of the detector can be amplified, registered, and is passed on to various evaluating devices.

2.2 Equipment of the Instruments

For the in situ evaluation of thin-layer plates, several modes of measuring are possible. Measurements in the reflectance mode and in the fluorescence mode are the most commonly used ones. Scanning in the transmission mode and simultaneous reflectance and transmission mode are also in use.

2.2.1 Reflectance Mode of Scanning

In the reflectance mode of scanning, the monochromatic light either falls at an acute or right angle to the plate; the diffusely reflected light is then detected. When the beam of light meets a sample spot, a part of the light is absorbed, and only the remaining reflected light is measured. The difference between the signal of a place free of substance and the signal of a spot is the real measuring signal. This can be registered after passing an amplifier.

As a source of light, deuterium lamps are in use for the ultraviolet (UV)-region and tungsten or halogen lamps for the visible region. Xenon lamps are applied both in the UV- and in the visible region. Moreover, a few lines of the spectrum from mercury lamps are used for measurements in the reflectance mode. For a very short wave length (<220 nm), Ebel and Geitz [5] recommend hollow cathode lamps as used in atomic absorption spectroscopy. Deuterium lamps have a continuous spectrum ranging from 200 (185 — Zeiss) — 350 nm.

Xenon lamps have a particularly high intensity of radiation and are therefore suitable for double-beam densitometers, as the beam is divided into a sample ray and a reference ray. The region of use ranges from approximately 220 nm to 700 nm.

Tungsten- and halogen lamps are generally used between 350–650 nm. For some lamps this range is expanded to 800 nm. Zeiss quote a range up to 2500 nm [6] for their tungsten lamps. With mercury lamps a few lines of the spectrum between 254 and 578 nm are used, some of which radiate intensely.

The monochromatic light is produced by means of quartz prism- or grating monochromators. In a few cases filters are still used. Quartz prism monochromators have a particularly good spectral dispersion in the UV-range. For a slit-width of 0.1 mm the band-width is 0.14 nm at a wave-length of 200 nm and 2 nm at 400 nm [6]. The field of use ranges from 185 to 2500 nm. With quartz prism monochromators, however, a considerable loss of light energy has to be taken into account. The loss of energy by grating monochromators is much less. These monochromators have a band width of 10 nm throughout the entire field of use ranging from about 200 to either 630 or 800 nm depending on the type.

When HPTLC-plates came into use, the incident beam of light had to take on smaller dimensions. This was achieved either by variable apertures or, in order to avoid a loss of energy, by a dimension reducing imaging. For this purpose some instruments have been equipped, in addition, with so-called micro-optics. These can be brought into the beam by a quick exchange of a light tube (Zeiss-densitometer), or by switching to another lense-system (e.g. Camag-densitometer).

Photomultipliers used within a range of 200 to 650 nm serve as detectors. Only recently broadband photomultipliers with a spectral sensitivity of 185–850 nm have appeared on the market [7]. In the range above 650 nm the photocell operating up to 1100 nm or the photoresistance responding up to 2500 nm are used. Occasionally attempts have been made to apply flame ionisation detectors to thin-layer chromatography [8–10]. In this case, however, it is necessary to carry out the chromatography on rods, or strips or tubes coated with adsorbents.

2.2.2 Transmission Mode of Scanning

In the transmission mode of scanning the beam of light passes through the plate vertically and comes to the detector which, in this case, is under the scanning table. As the adsorbent layers and the glass plates are not permeable for short wave length light, the transmission mode of scanning is only applied in the visible wave length range. The corresponding sources of light and the detectors are the same as those for the reflectance mode of scanning.

Because of the limitation on the visible range, and because of the less favorable signal-to-noise ratio in comparison to the reflectance mode of scanning, not all instruments are equipped for the transmission mode.

2.2.3 Simultaneous Reflectance and Transmission Mode of Scanning

For the simultaneous reflectance and transmission mode of scanning the equipment for measuring is limited to the visible wave length range and requires a special device. This device brings together the signals from the two detectors for the reflectance- and for the transmission mode of scanning; the relation of factors of the two signals can be continuously adjusted. The relation of factors depends on the inclination of the baselines obtained for reflectance and transmission; it is chosen such that baseline variations can be almost eliminated.

2.2.4 Fluorescence Mode of Scanning

In the fluorescence mode of scanning the path of the light normally corresponds to the one in the reflectance mode. The light which excitates fluorescence passes a monochromator or a corresponding filter and falls onto the plate. The emitted light comes to the detector above the plate. The diffusely reflected light of the excitation wave length is retained by a cut-off-filter. Cut-off-filters only allow light above a certain wave length to pass; thus they filter out the light of the excitation wave length; this is mostly shorter than the wave length of the emitted light.

If the emitted light is strong enough, monochromatic filters can be applied; this allows a more specific selection from the spectrum of the emitted light. Moreover, monochromators allow an even more specific selection of the emission wave length. Here, too, the emitted light must be intensive enough because energy is lost while passing a monochromator.

In very few cases the path of light in the fluorescence mode of scanning is chosen vice versa (possible with the instrument from Zeiss [6)]). The lamp is placed on the scanning head, and after passing monochromatic filters the light reaches the plate at an angle of 45°. The fluorescence light passes a monochromator and is conducted to the detector [11)].

Either mercury lamps, mostly of the medium pressure or the high pressure type, or xenon lamps serve as a source of light. The xenon lamp is regarded as a quasi-continuum lamp and has the advantage that it can be used in a wider range of the spectrum. Only several lines of the mercury spectrum, on the other hand, provide enough energy and thus the wave length of excitation often cannot be adjusted specifically enough to the substances to be measured. However, the intensity fluctuations of a mercury lamp are considered less than those of a xenon lamp [12)].

Generally the scanning procedure in the fluorescence mode is the same as in the other modes of scanning; the chromatogram is scanned by a slit-beam or point-beam. As the intensity of fluorescence of a sample spot is directly proportional to the quantity of the substance of the spot, the entire spot fluorescence can also be measured at one time. For this an iris diaphragm is used, the diameter of which is chosen somewhat wider than the diameter of the spot to be measured [12)].

3 Evaluation Systems with Different Kinds of Scanning

According to the measuring procedure two kinds of densitometers can be distinguished, slit-beam densitometers and point scanning densitometers. With some of the instruments which are on the market both measuring procedures are possible and can be selected by means of a switch.

3.1 Slit-beam Densitometers

A slit shaped beam brings light on to the chromatoplate. The plate lies on a table which moves with an adjustable constant velocity linearly under the slit. This procedure is also called linear scan in the literature. For circular or anticircular developed plates a turntable is used. The scanning procedure is performed radially or a peripheral measurement is carried out with the slit positioned in the direction of measurement or perpendicular to it (Fig. 2).

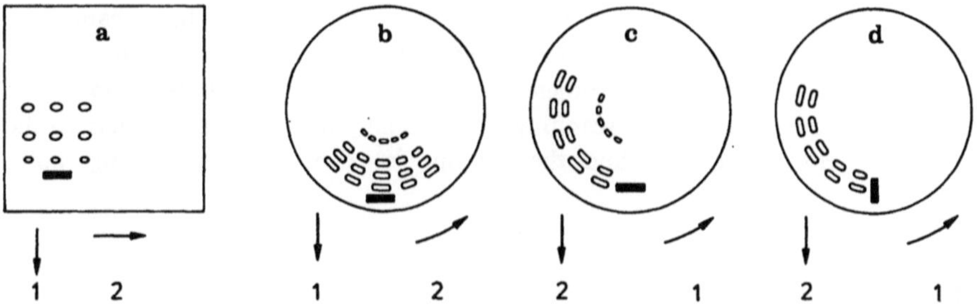

Fig. 2. Slit-beam evaluation scheme 2a: linear scan, 2b radial scan, 2c and 2d peripheral scan;
1 = direction of the plate table
2 = direction of track-changing

The width and length (height) of the slit is adjustable by a variable aperture. The slit can be optically imaged on the plate. If sample spots are poorly separated, a small slitwidth is chosen; this enables the spots to be measured separately. In some instruments the band-width of the monochromatic light is actually determined by the slit-width, and thus, in order to maintain good optical resolution the latter should be kept small.

The length of the slit (track-width) is adapted to the width of the sample spot to be measured. If the samples have been applied as lines, the length of the slit is chosen according to the part of the line-shaped chromatogram which is to be measured.

Slit-beam densitometers are widely spread and have been brought onto the market by several producers such as Zeiss, Schoeffel, Camag, Shimadzu and Farrand. Results obtained from these instruments are good and the plates can be measured in a relatively short time.

The following aspects, however, have to be taken into account:

1) The light intensity over the whole slit is not quite uniform. Therefore it is very important to position the spots to be measured exactly under the slit.
2) In chromatography great care must be taken in achieving uniform spots. Irregular spots cannot be measured precisely by the slit-beam densitometer.
3) The area values measured depend on the scanning direction. For this reason all samples and standards on a plate are measured in one direction, mostly in the direction of chromatography.

3.2 Point Scanning Densitometers

In these densitometers, a small bundle of light images a square or round spot on the thin-layer plate.

This mode of point scanning does not require any positioning. In general, the linearity of the calibration curves is better than for those obtained from slit-beam densitometers. The accuracy of scanning is independent of the shape of the sample spots; even irregularly formed spots can be measured exactly. The direction of scanning has no influence upon the measuring results, as experimentally demonstrated by Herold [13]. One disadvantage, however, is that the spot-beam cannot be chosen smaller than 0.2 mm [14] due to the plate structure. The signal-to-noise-ratio is poorer than with slit-beam densitometers. Furthermore discussions arose concerning the baseline correction [15]. The scanning point usually moves perpendicular to the direction of chromatography and the baseline is taken near the turning point

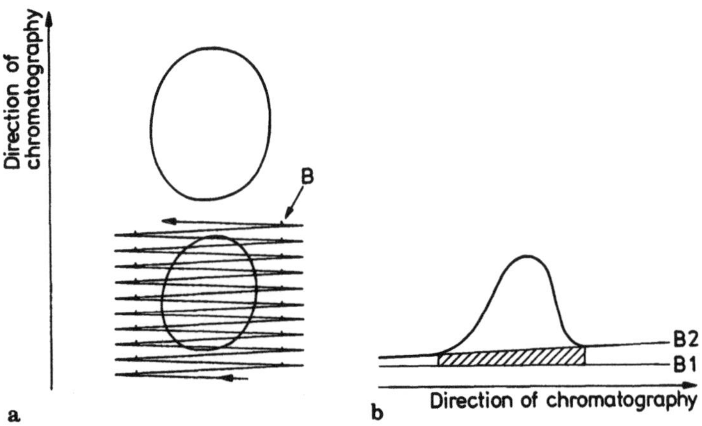

Fig. 3. Spot beam evaluation scheme; B = Baseline measuring; B1 = background of the plate, B2 = background of the plate + sample matrix

(Fig. 3a). In this case the baseline correction is confined to the plate background directly adjacent to the sample spot. The background caused by the sample matrix is not taken into consideration (Fig. 3b). This can lead to excessive errors as the measuring signal can be greatly increased.

According to the type of instrument in use the scanning point describes a sinusoidal-("flying-spot" system), a zigzag- or a meander path on the plate.

3.2.1 "Flying-spot" Scanning

In the "flying-spot" system [16] the plate is directed sideways by an excenter, perpendicular to the movement of the spot-beam on the plate.

It has been assumed that the movement correlates to a sine function and the velocity to a cosine function. As this would lead to considerable integration errors the instrument referred to has been provided with a cosine corrector; this corrects the measuring signal coming from the photomultiplier after it has been amplified. According to Ebel et al. [17], however, the velocity of the oscillating movement corresponds to an entirely different function (formula and figure can be found in the cited literature); thus there is no guarantee that the flying-spot scanner exactly measures each part of a sample spot even once.

3.2.2 Zig-zag Scanning

In terms of these instruments the plate table is moved longitudinally and perpendicularly such that the movement corresponds to a triangular function [18]. The square spotbeam is adapted to the velocity of the scanning stage so that every part of a sample zone is exactly scanned twice [13,17,19]. According to Ebel et al. [17], however, the two lateral turning points should not be registered, because the chromatoplate comes to a short standstill.

3.2.3 Meander Scanning

The meander shaped scanning of the chromatograms has been realized by a computer-controlled instrument (look part 5.2.1). The computer directs the scanning table so that the sample spots are scanned in the meander mode in the direction of the chromatography.

By measuring in the direction of the chromatography an exact baseline correction is possible, because the baseline is controlled before and after the sample spots and not laterally.

4 Instruments with Different Photometric Systems

4.1 Single-beam Densitometers

As with spectrophotometric measurement in solutions, single or double-beam instruments can be used for the in situ scanning of thin-layer chromatograms. When scanning with a single-beam densitometer, irregularities of the adsorbent layer (varying thickness of the adsorbent, different particle-size or density) and irregularities by impurities e.g. by the solvent front, interfere with the results.

4.2 Double-beam Densitometers

Using double-beam scanning one attempts to eliminate irregularities caused by the adsorbent layer or by contamination. The monochromatic light is divided into a sample beam and a reference beam. On the chromatoplate every other track is kept free of sample; this free track serves as a reference.

Sample- and reference tracks are scanned at the same time. The reflected light meets two different photomultipliers. Both signals are separately amplified. Because of the immediate ratioing, only the signal caused by the sample is registered. Near the solvent front the background can be increased. Double-beam scanning can eliminate interferences caused by both solvent and β-fronts.

The producers of the instruments have emphasized that, with the double-beam instruments, a higher degree of accuracy is achieved. Ebel and Kussmaul [20] have tested, in numerous measurements, how great the gain in accuracy really is. They have simulated a double-beam densitometer on a single-beam instrument by means of a computer. They obtained relative approximated standard deviations of 3–5% for single-beam- and 2–4% for double-beam densitometers. Since, meanwhile, the quality of the commercial thin-layer plates has been essentially improved, the differences today may be even smaller.

It is, however, disadvantageous that, with this procedure fewer samples can be applied on the plate, as every second track must remain free for reference. Furthermore sample and reference are, in fact, scanned at the same time, but not at the same place. In many cases the adsorbent layer of two directly neighbouring tracks will be very similar, but impurities on, or damage to, one of the tracks can lead to considerable errors.

4.3 Dual-wavelength Densitometers

Because of the disadvantages of the double-beam instruments attempts have been made to scan both simultaneously and at the same place. This is possible with dual-wavelength densitometers. A sample beam and a reference beam are generated by two monochromators. The sample wavelength is chosen in the absorbance range of the sample, the reference wavelength in a range where the sample shows no absorbance. Sample- and reference beam meet the track in a quick alternating sequence. Sample- and reference wavelength must be close to each other, as the scattering coefficient depends on the wavelength. For wide spectral curves of the samples, therefore, scanning cannot take place at the absorbance maximum; a wavelength must be chosen not too far from the wavelength in the absorbance-free field, i.e. on the rising or falling part of the absorbance curve. This means a loss of intensity of the measuring signal.

Interference caused by the solvent front or by a β-front is not eliminated, as these fronts do not have the same absorbance at different wave lengths.

For dual-wavelength densitometers, Ebel and Kussmaul [20] have found a relative standard deviation of 2.5–4%.

4.4 Additional Device for Simultaneously Measuring Reflectance and Transmission

A further possibility to measure both at the same time and at the same place and additionally with the same wavelength is given by the simultaneous reflectance and transmission scanning.

Treiber et al. and Stöllnberger [21] have described the principle of this kind of measuring. They realized that the baselines in the reflectance and transmission mode run in opposite directions and that they appear as mirror images if the amplifier is adjusted accordingly; signals originating from the absorbance of a sample, on the other hand, run in the same direction. A special device has been developed [22] with which the ratio of the factors for the signals of the reflectance- and transmission mode can be adjusted continuously. The two signals are brought together and the resulting signal is passed on to the indicating instrument. Thus irregularities of the adsorbent layer can be eliminated, interferences originating from the absorbance like fronts, however, cannot.

Another positive effect is that the measuring signal of the samples is increased since it is the sum out of the reflectance and the transmission signal. According to Jork [23], by this procedure the detection limit can be lowered by the factor 10 to 100.

In spite of the theoretical significance of this kind of measuring it has not been frequently put into practice. Unfortunately this method can only be applied to the visible wavelength range; the adsorbent layers do not allow transmission measurements in the UV range.

For routine runs, single and double-beam instruments are used frequently; dual wavelength densitometers are primarily applied in the field of research. One scanner on the market can work both as a double-beam and a dual wavelength instrument; the systems can be changed by means of a shutter plate [24].

5 Automatic Systems

The measuring procedure in quantitative thin-layer chromatography is time consuming, tiring and prone to personnel errors; it is also labour intensive, since it is common practice to apply several spots of the same sample and to make the mean value of the measurements in order to eliminate uncertainties resulting from irregularities of the adsorbent layer. Therefore, relatively early on in the development of these instruments attempts were made to automate the measuring procedure.

Meanwhile, two systems have been developed, i.e., automatic scanning instruments and computer controlled systems.

5.1 Automatic Scanning Instruments

The simplest automatic scanning instruments perform an automatic track change. The chromatoplate is positioned on the first measuring track, then the entire given track is scanned. After measuring a track the scanning table goes back to the starting position and runs automatically to the next measuring track corresponding to the previously chosen track distance. The total number of the tracks to be scanned is also given in advance. This method has two main advantages; the one being that the procedure is very fast and the other that even spots which are not completely separated can be measured. Today a number of scanners are already equipped with such track-changing devices (e.g. Camag, Shimadzu scanner). In split-beam densitometers, however, not necessarily all spots on a track are registered in the best possible way. This is due to the fact that not all the spots always lie exactly on the track and, in particular, that spots which have run further up or which lie close to the edge of the plate can be slightly off the track. This can lead to considerable errors, as slit-beam densitometers require best possible positioning in the measuring slit.

Less difficult, however, is the use of a track-changing device in connection with a zig-zag scan. One has to choose the oscillation width big enough to completely register all spots deviating slightly from the track.

In order to eliminate the disadvantage, automatic devices have been developed for slit-beam densitometers which can scan a track several times [25, 26]. Distance and number of scan courses as well as the distance of the sample tracks are entered into the automatic instrument in advance. The highest value derived from scanning a track repeatedly is adopted as measuring value for each spot. This method of scanning is, unfortunately, time consuming.

Particularly for the Zeiss KM3 an automatic apparatus was developed which enables the scanner to position each spot on the track separately and thus guarantees that all spots lie in the scanning slit in an optimum way [27]. The latter is a microprocessor controlled instrument. The movement of the scanning table is brought about by stepping motors. Plate specific parameters, such as track length, track distance and number of tracks as well as peak realizing criteria like step width in the course of the peak search, peak sensitivity and slope criteria are adjustable by a thumb wheel. If during the course of the search a peak is realized, the positioning is performed while going back. Then comes the scanning course itself; the spot is integrated and corrected by subtracting the baseline. One disadvantage is that the relevant spots cannot be directly aimed at, for example, by bridging the gap between two spots to be measured or by running according to Rf-values. Thus all additional spots on a sample track which are not of interest are registered too, if they surpass the peak sensitivity value. The results obtained are then printed out. A further automatic calculation and a documentation of the results requires an additional connection to a computer.

Automatic scanning instruments are usually coupled to integrators or the integration is carried out by built-in microprocessors. With computer integrators, the user is tied to the application scheme provided by the integrator (arrangement of standards and samples on the plate). At times this scheme is not the most suitable one for the thin-layer chromatography.

Programming of microprocessor-controlled instruments has so far not been possible; this makes the systems inflexible.

As a point of particular interest, the so-called linearizer may be mentioned [18]. Three linearizing functions can be chosen by means of a switch; linearization of the calibration curve is then possible.

Final calculations and documentation of the results require the connection to a computer.

Measurements are mostly accompanied by a recorder. When measuring automatically such an additional possibility of control is important. Zig-zag scanners can also be accompanied by a recorder during the quantitative evaluation, as only the peak value of each swing is recorded [18].

5.2 Fully Automatic, Computer-controlled Systems

Two fully automatic, computer-controlled systems are known today: that of the Ebel group and that of the Kaiser group. Originally, Ebel and collaborators dealt, in particular, with scanning and positioning. Later on a number of evaluation methods

The spots to be measured can be attended one after the other; they can also be aimed at directly by Rf-values [38]. Thus additional spots are not registered unnecessarily. Measurements are performed by means of peak height, area or peak approximation. For overlapping spots peak approximation [15, 39], derivatives of the reflectance curve [15, 40] or dual-wavelength technique are recommended [38, 41]. For irregularly shaped spots a meander scanning with a small measuring area is chosen. The scanning is done in the direction of the chromatography, so that the baseline before and behind the spot can be measured (Fig. 5). As a measuring value the peak area or peak height is taken.

The linscan was developed particularly for chromatograms where samples are applied as lines, or for plates with a concentrating zone; the latter also give a band-shaped chromatogram. After realizing a substance zone in the direction of the chromatography (y-direction) the borders of the zone are stated in x-direction. Then the zone is measured in a meander-shaped way in y-direction, so that the baseline before and behind the zone can be measured (Fig. 6). The results can be expressed as a sum of the areas or the heights.

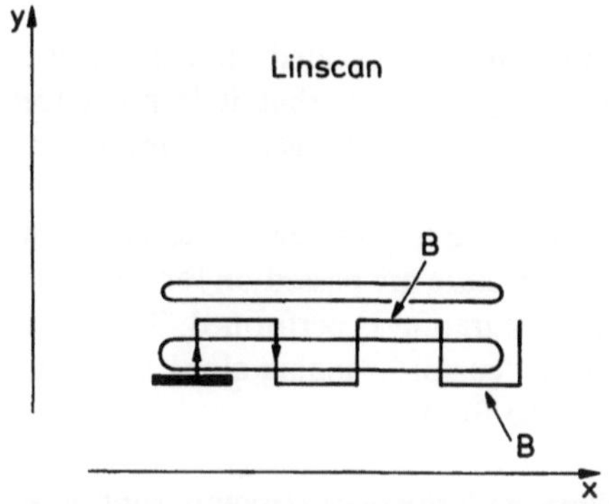

Fig. 6. Meander scanning for band-shaped chromatograms; B = Baseline measuring

The evaluation is possible by means of external or internal standards. Various schemes of sample application are assigned for that, among others the application scheme according to the data-pair-technique [42]. The system is easily adapted to any other application scheme.

As in thin-layer chromatography the calibration curves are usually only linear in a limited range, various linearisation functions are provided in the software program of this system, but also non-linear calibration curves for working in a wider concentration range [15].

If for a scanner a monochromator drive is disposable and controllable by a computer, like for example with the instrument KM 3 Zeiss, spectra and their derivatives can be recorded. A spectrum can be obtained by taking the difference between substance- and background spectrum after a logarithmic transformation or a transformation by the Kubelka-Munk-function.

A computer provides a complete analysis report. Using these reports even the course of the automatic measurement can be controlled.

5.2.2 System of the Kaiser Group

Recently Kaiser and collaborators have developed new possibilities by means of a computer-controlled system [43,44]. They have coupled a desk computer (Apple II) via an AD-converter to the signal output of a Camag-scanner. Additionally, the computer is connected to the control electronics of the scanning table of the densitometer. Thus the transmission of the raw measuring data and the computer-control of the automatic scan is possible.

While developing their system the authors presumed that the background structure of the thin-layer plate significantly falsifies the measuring signals. They therefore emphasize that an integration may deliver false results, unless it is performed in a very high concentration range. By subtraction of the plate structure from the raw data and by computer integration the authors achieve a reproducibility and accuracy in average analyses of 0.1 %.

If the plate structure is to be subtracted from the raw data, the plates are measured precisely on those tracks on which later on chromatography is performed. Here it is necessary for the plates on the scanning table to be repeatedly positioned very precisely. This method, however, only works in connection with the anticircular-technique [45]. The extensive data material obtained is stored on so-called "floppy discs".

The process of precleaning the plates and carrying out repeated measurements is work-intensive, but this is partly made up for by the fact, that it is no longer necessary to apply several spots of a sample, as there is no longer any uncertainty resulting from the irregular plate structure.

The data stored on floppy discs are easily accessible for further processing. Thus the data cleared after a subtraction of the plate structure are passed on for computer integration. In difficult cases a so-called video-integration is performed. The signal curve is shown on a video screen. By means of a "cursor" running along the signal curve, beginning and end of the integration can be fixed.

Kaiser maintains that, contrary to the integration by electronic integrators, only the computer integration of the chromatogram signal with the plate structure subtracted delivers accurate results in thin-layer chromatography.

6 Performance of the Densitometers and Comparison with Instruments of Other Kinds of Chromatography

6.1 Advantages and Disadvantages of the "Off-line" System in Thin-layer Chromatography

If one tries to evaluate the performance of the densitometers in comparison with other types of measuring instruments used in chromatography, the following must be taken into consideration:

In thin-layer chromatography the separation of a mixture of substances and the measurement of the separate spots are fully independent processes. Jänchen [46] talks of it as an "off-line" technique in contrast to column chromatography, which he calls an "on-line" technique.

This "off- line" technique has the following advantages:

1) The measuring procedure can be repeated as required. By repeated measuring a higher degree of accuracy is achieved. It is even possible to carry out the repeated measuring with only some of the spots on a chromatogram.
   The measuring procedure can be optimized without further chromatography. It can be changed as required, for instance by choosing different measuring modes, such as reflectance or fluorescence mode of scanning.
   This possibility is, however, limited if substances are to be measured which are quite sensitive to light or oxygen or which, by some other means, undergo rapid changes on the plate.
   The various spots of a chromatogram can be scanned in different ways. A part of the chromatogram, for example, can be measured in the reflectance mode, possibly at different wave-lengths, the other part in the fluorescence mode or after derivatization, as shown by Rücker et al. [47]. All this is possible after only one separation procedure. In column chromatography it is necessary to repeat the separation step for each of these procedures, as the separating and measuring processes are combined.
2) The chromatographic system can be changed quickly and there are many possibilities of variation. The measuring system is rapidly adjustable and thus immediately adaptable to various analytical tasks. This way several different analyses of thin-layer chromatography can be performed on one day.
   In column chromatography the changing and conditioning of columns and the changing of the mobile phase are time consuming, and therefore it is less suitable when dealing with constantly changing analyses.
3) Thin-layer plates are only used once, while in column chromatography the stationary phase is used repeatedly. Therefore even samples highly contaminated by the matrix can be directly applied on the plate without being cleaned up. In column chromatography an irreversible contamination of the stationary phase has to be taken into consideration.
   From the numerous examples in the literature two might be quoted here:
   Amin and Jakobs [48] dissolve suppositories and ointments, which contain several active ingredients (hydrocortisone or its esters, an antihistaminic, a local anaesthetic and a disinfectant), in chloroform/methanol. They apply the solution directly on a thin-layer plate without freeing it from the suppository or ointment base. After the separation the active ingredients are scanned with variation coefficients of 1.29–3.56%.
   Grebian et al. [49] apply biological material directly (urine) or after adding methanol and centrifuging (plasma, faecal samples) on thin-layer plates. After the chromatography the pteridines triamterene, hydroxytriamterene and its sulphate ester are measured. The average variation coefficient with 20 ng/spot from urine is 2.9%, with smaller concentrations it is higher; with plasma it is indicated as 6%, with faecal samples it has not been noted. The recovery from urine and plasma is 100%, from faecal samples 95%.
4) The number of mobile phases to choose from is great, since the solvents are evaporated after chromatography and no interference with the measurement can

occur. In column chromatography the mobile phase has to be adapted to the detector, thus, for example, in HPLC when applying UV absorbance many solvents cannot be used because of their self absorption.

5) After the chromatography derivatization is possible in a simple way. All samples on one plate are derivatized simultaneously. By spraying, dipping or exposing to vapours the samples on the plate can be changed, stained or transformed into fluorescing compounds and so become more easily measurable. Here two examples from the field of pharmacy may be quoted [47, 50] as well as an article about postchromatographic derivatization with numerous references in pharmacokinetic and drug metabolism studies [51]. Kreuzig [52] published a review about the techniques of derivatization in quantitative thin-layer chromatography.

This "off-line" system, however, has some disadvantages, too:

1) The method cannot be fully automated.
   Automatic sample applicators and automatic scanners exist. The changing of the plates in the scanner is done manually. A complete automatization would be difficult and expensive and will therefore not be realized in the near future.
2) Calibration functions cannot be transferred from one plate to another, and so a new calibration must be carried out for each plate.

6.2 Sensitivity

In thin-layer chromatography very small amounts of substances are measurable. Quantities down to the lower nanogram range can still be evaluated. When scanning in the fluorescence mode even amounts in the lower picogram range have been determined [3]. The measuring range for the absorbance mode is indicated as 10 to 5000 ng, for the fluorescence mode as about 0.1 to 500 ng [3]. But it may not be overlooked, that only small volumes of substance solutions can be applied on the plates. For quantitative analysis, when applying the solutions as spots, the optimum volume for standard plates is 0.5–2 µl, for HPTLC-plates 200 nl. Therefore the concentration of the solutions must be correspondingly high.

The measuring instruments in high performance liquid chromatography reach a higher sensitivity. Moreover bigger volumes are injected, usually 5–20 µl, so more dilute solutions of the samples can be applied. Furthermore the sensitivity of the detectors has recently been increased significantly.

Verpoorte and Svendson [53] have compared thin-layer chromatography and high performance liquid chromatography when analysing alkaloids; they found the detection limit in HPLC to be 1–100 ng and in TLC to be 10–100 ng.

The gas chromatography on packed columns and using flame ionisation detectors is, in general, comparable to thin-layer chromatography; the differences for specific substances can be great, depending on the response to the detection method. Small volumes (1–2 µl) are injected, so that also here the concentration of the sample solutions has to be correspondingly high. In capillary gas chromatography much smaller quantities are detected. Once again concentrations and volumes are injected as given above; due to a splitting system, however, only a small fraction actually reaches the column.

In thin-layer chromatography Kaiser et al. [43, 44] achieve an enhancement of sensitivity by subtracting the plate structure from the measuring signal. This way they improve the detection limit by a factor of about 10.

6.3 Accuracy and Reproducibility

When scanning in the absorbance mode the total error is, according to Jork [54], ±2–4% (variation coefficient), in the fluorescence mode < ±1%. Grimm [55] has calculated a mean variation coefficient of ±2.5% from numerous data in the literature and from his own analyses; the variations lie between 0.7 and 5.2%; the statistical data relate to the total analysis with manual evaluation.

The total error consists of several parameters, in which the error of the scanning instrument takes the smallest part. Jork [54] has investigated the error of the measuring instruments leaving out the adjustment error on three different evaluating units (scanner + integrator) and has found variation coefficients of ±0.16–0.44%. He indicates the manual adjustment error as 0.25–0.46%. When performing numerous analyses, however, the manual adjustment error increases due to fatigue of the working personnel. In automatic scanning instruments, where the adjustment is only performed once at the beginning of a plate and with which then the scanning goes on according to preselected distances, the adjustment error is greater than the error by manual adjusting. The error becomes smaller with an automatic system; this positions each spot separately. As already mentioned, with the system of the Ebel group an accuracy of <1% is reached and after an optimal chromatography even an accuracy of 0.3–0.5%; the reproducibility, or strictly speaking, the error of the measurement of the same lane with appropriate positioning is indicated according to the different evaluation methods and lies between 0.42 and 0.56% for peak height, between 0.49 and 0.69% for the numerically integrated area and between 0.27–0.48 or 0.49–0.6% for peak height or area by approximation using a Gaussian function [38].
manual adjusting. The error becomes smaller with an automatic system; this positions

The somewhat higher errors observed when evaluating by area depend on the difficulty in exactly determining the beginning and end of integration.

With the previously described system of the Kaiser group, characterized by subtraction of the plate structure from the measuring signal and by computer integration, a reproducibility and accuracy of about 0.1% is reached when performing average analyses [43].

From these results it can be seen, that with the most recently developed measuring instruments, a very high degree of precision, comparable to that of high performance liquid chromatography and capillary gas chromatography, is achieved. If a large number of samples is to be analysed, then automation is recommended; for slit-beam scanners the individual positioning of each spot is essential. If the variation coefficient is too high, the error will occur in one or both of the analysis steps which are independent of the evaluating system, i.e. the sample application or the chromatography.

6.4 Wear and Maintenance

Densitometers are very well suited for routine use. They are highly reliable, even when used permanently. The need of maintenance and attendance is extremely low.

In this respect the thin-layer chromatography is superior to other kinds of chromatography, above all to high performance liquid chromatography.

The instruments do not have to be cleaned regularly. Now and then optical parts, which are exposed to contaminated laboratory air, have to be cleaned or replaced. From time to time a new lamp must be inserted. Tungsten and deuterium lamps are highly durable, and can often be used for much longer than a year, if they are not contaminated by laboratory air. Xenon and mercury lamps, however, are not quite as stable and have to be replaced more often.

When using other kinds of chromatography, it is often necessary to clean and to replace the column. The thin-layer plate, however, is independent of the measuring instrument and is used only once. Both the sample application system and the transport of the mobile phase are independent of the densitometer. The latter can cause many problems and require a lot of maintenance particularly in high performance liquid chromatography.

6.5 Time

Detailed information about the requirements of time for analyses in thin-layer chromatography are given by Jänchen [46,56]. A table [46] is given showing the time required for each individual step of analysis, such as sample application, chromatographic development and quantitative scanning. The total time depends on the number of samples per plate, that is the number of samples to be analyzed simultaneously. When using HPTLC-plates the total time per sample, beginning with the sample application, is indicated as 1.3–2 min. When taking into consideration the length of time for calibration (this being necessary for each separate plate), the following times are calculated for the duplicate determination of an unknown sample: 6 calibration tracks and a total number of 24–32 tracks <4 minutes, 6 calibrations and a total number of 16 tracks 6.3 minutes.

Very little time is required for sample preparation; excessive clean-up of the samples is not necessary.

The time for the chromatographic development is indicated by Jänchen as 4 and 2 minutes for the circular and anticircular technique and as 12 for the linear development on HPTLC-plates. Standard plates require about twice as much time. The exact time will of course depend on the developing system. The developing distance, however, cannot be changed essentially. In gas chromatography and high performance liquid chromatography the developing distance is more variable. The developing times vary correspondingly. The separating efficiency is thus better than in thin-layer chromatography.

Compared to other kinds of chromatography the measuring time in thin-layer chromatography is very short, as the substances do not have to be eluted. Multiple measurements of calibrations on each plate, however, are necessary; this reduces the advantage slightly.

The measuring time depends on the measuring distance. Moreover it is decisive whether each spot is positioned separately or whether the positioning is performed only once at the beginning of a plate or of each track. For manual scanning Zeiss [6] indicates a measuring time including instrument adjustment of <3 min, when measuring 10 spots, and of <5 min, when scanning 10 tracks of 10 cm. The scanning times given by Jänchen [46)] relate to the Camag scanner with built-in scanning automation, which performs an automatic track-change without repositioning. Scanning half of the separation distance is said to take about 13–18 min for 24 tracks. When positioning each spot before measuring it the scanning time additionally depends on the number of spots per track. A less modern measuring system takes 45–55 min for 18 tracks and two spots per track [34, 35)], the new system of Ebel [15)], on the other hand, may work about 3 to 4 times faster.

Very little time is required for maintenance and cleaning the instruments. Plates do not need to be cleaned as they are used only once. In contrast column flushing in high performance liquid chromatography and column cleaning by heating in gas chromatography are both time consuming.

Densitometers are ready for use for quantitative measurements about 10 min after switching them on.

6.6 Personnel and Costs

Personnel requirements depend on whether the sample application and the measurement is done manually or automatically. It is still common practice to apply the samples manually with microcaps or nanocapillaries and the automatic devices have not yet reached the same accuracy and speed as the manual application.

A skilled laboratory assistant can apply 20 samples in <10 min. The scanning, however, is done more and more automatically. Jänchen [56)] demonstrates, how much time is consumed by the various analysis steps and how much of this time is taken up by the laboratory assistant when sample application and measurement are performed automatically. When 33 samples are analyzed twice he gives a total analysis time of 148 min, of which 30 min are the assistant's time, that is 1 min for 1 sample. This does not include the time for sample preparation, as the latter is dependent on the type of sample. In general, sample preparation requires less time in thin-layer chromatography than in other kinds of chromatography.

As already mentioned, the densitometers require little maintenance and cleaning unlike the instruments in gas and high performance liquid chromatography. This is advantageous as less personnel is needed. Laboratory assistants working in thin-layer chromatography have to be able to work precisely, but they do not need a special technical knowledge.

The low personnel and maintenance requirements are beneficial as far as costs are concerned. The price of the densitometer will depend on the degree of automation. As far as operating costs are concerned plates and solvents are the most important factors. Occasionally a new lamp is required for the densitometer.

6.7 Fields of Use

A literature survey on the fields of use of quantitative thin-layer chromatography was carried out by Jork and Wimmer [54)]; the first part of the survey appeared in 1982 and is successively being added to. In this collection of operational methods and literature, publications are quoted from the fields of pharmacy and pharmacognosy, clinical chemistry, environment analysis, analysis of food and cosmetic articles, and analysis in industry and technical sciences.

Further extensive collections of literature by producers of plates and instruments give information on the possible use of quantitative thin-layer chromatography.

Chromatographic methods are mainly used in organic analytical chemistry but they have also been proved suitable for inorganic analyses [57]. Here thin-layer chromatography has the advantage that it offers several possibilities of reactions for the detection after the chromatography. In high performance liquid chromatography derivatization after the separation requires additional devices and is more restricted in comparison to thin-layer chromatography.

Chromatographic methods are specific. They thus find a wide range of use especially in the field of pharmacy, where in particular stability tests require the separation of decomposition products. Grimm and Schepky [58] describe the in situ evaluation of thin-layer chromatograms as one of the most important methods of analysis in stability tests.

High performance liquid chromatography is the method which competes most with thin-layer chromatography. Verpoorte and Baerheim Svendsen compare the possibilities and limits of both methods in the analysis of alkaloids [53]. According to them the advantages of quantitative thin-layer chromatography are, that the clean-up of the samples is not so important as in high performance liquid chromatography and that several samples can be analyzed simultaneously. High performance liquid chromatography leads to a better separation, the risk of sample decomposition is smaller and the sensitivity is higher than in thin-layer chromatography.

Personal experience has shown quantitative thin-layer chromatography to be suitable for quickly changing analyses, furthermore in routine runs, where sample preparation steps can be omitted or where derivatization is necessary for detection. However, it is less suitable for substances which are very sensitive to oxygen or light and not suitable for volatile substances.

7 Conclusions

A system is as weak as its weakest link. In thin-layer chromatography today the densitometers are not anywhere near the weakest link in the "off-line" system.

At present attempts are being made to optimize the chromatography, and a great deal of faith has been placed in OPTLC (overpressured thin-layer chromatography). With this technique an extension of the development distance, better resolution and a shorter separation time are achieved. In 1982 a report was published on the applicability of this technique in the analysis of various substance groups [59].

OPTLC on the other hand requires plate optimization and sorbent layers with particularly fine particles are necessary.

The sensitivity of the densitometers has been essentially improved in recent years. Their light energy has been intensified; this was made possible by short light paths, small losses of energy resulting from better monochromators and micro optics.

The noise of the instruments has been reduced. As far as detectors are concerned the development, however, has been slow. More sensitive and specific detectors might be desired.

By automation and connection to a computer the quantitative evaluation in thin-layer chromatography has become more reliable, faster, more versatile, and has achieved a greater degree of flexibility, accuracy and sensitivity.

8 Acknowledgements

The author thanks Mrs. Gillian Bishop-Freudling for "polishing" the paper with regard to language problems.

9 References

1. Leaflets of the firm Camag concerning the device Eluchrom
2. Dibbern, H. W., Wirbitzki, E.: GIT Fachz. Lab. *19*, 117 (1975)
3. Hezel, U.: Kontakte (Merck) *3*, 16 (1977)
4. Ripphahn, J., Halpaap, H.: J. Chromatography *112*, 81 (1975)
5. Ebel, S., Geitz, E.: Kontakte (Merck) *1*, 44 (1981)
6. Leaflets of the firm Zeiss, Oberkochen, concerning the instrument KM 3
7. Leaflets of the firm Camag concerning the TLC/HPTLC Scanner 76510
8. Mukherjee, K. D., Spaans, H., Haahti, E.: J. Chromatogr. *61*, 317 (1971)
9. Mukherjee, K. D., Mangold, H. K.: ibid. *82*, 121 (1973)
10. Mukherjee, K. D.: ibid. *96*, 242 (1974)
11. Hezel, U.: Angew. Chem. *85*, 334 (1973)
12. Hezel, U.: GIT Fachz. Lab. *8*, 694 (1977)
13. Herold, G.: Dissertation, Marburg (1975)
14. Jänchen, D. E.: Discussion during the Second Int. Sympos. on Instrumental HPTLC 1982
15. Ebel, S., Geitz, E., Hocke, J.: GIT Fachz. Lab. *24*, 660 (1980)
16. Leaflets of the firm Vitatron
17. Ebel, S., Geitz, E., Klarner, D.: Kontakte (Merck) *2*, 12 (1980)
18. Leaflets of the firm Shimadzu
19. Yamamoto, H.: J. Chromatogr. *116*, 29 (1976)
20. Ebel, S., Kussmaul, H.: Fresenius Z. Anal. Chem. *269*, 10 (1974)
21. Treiber, L. R. et al.: J. Chromatogr. *63*, 211 (1971)
22. Leaflets of the firm Zeiss, Oberkochen
23. Jork, H.: J. Chromatogr. *82*, 85 (1973)
24. Leaflets of the firm Shimadzu concerning the instrument CS-910
25. Christiansen, C. P.: Acta Pharm. Technol. Suppl. *7*, 131 (1979)
26. Leaflets of the firm Zeiss, Zürich
27. Leaflets of the firm Eckelmann, Wiesbaden-Schierstein
28. Ebel, S., Hocke, J.: Fresenius Z. Anal. Chem. *294*, 16 (1979)
29. Ebel, S. et al.: Kontakte (Merck) *1*, 39 (1982)
30. Ebel, S., Herold, G., Hocke, J.: Symp. "Quantitative thin-layer chromatography", Merck, Darmstadt 1975
31. Ebel, S., Hocke, J.: J. Chromatogr. *126*, 449 (1976)
32. Hocke, J.: Dissertation, Marburg (1976)
33. Ebel, S., Hocke, J.: Chromatographia *10*, 123 (1977)
34. Böhrer, I., Schmidt, P.: Acta Pharm. Technol. Suppl. *7*, 135 (1979)
35. Böhrer, I.: Kontakte (Merck) *1*, 19 (1981)
36. Ebel, S. et al.: Symp. "Quantitative thin-layer chromatography" Merck, Darmstadt 1979
37. Ebel, S., Geitz, E.: Fresenius Z. Anal. Chem. *296*, 369 (1979)
38. Ebel, S., Geitz, E., Hocke, J.: Instrumental HPTLC (Kaiser, R. E. et al. Ed.), Hüthig-Verlag, Heidelberg, Basel, New York 1980
39. Kaal, M.: Dissertation, Marburg (1979)

40. Ebel, S.: Instrumental High Performance Thin-layer Chromatography, (Kaiser, R. E. Ed.), Bad Dürkheim 1982
41. Ebel, S., Herold, G.: Chromatographia *9*, 41 (1976)
42. Bethke, H., Santi, W., Frei, R. W.: J. Chromatogr. Sci. *12*, 392 (1974)
43. Kaiser, R. E., Prošek, M. G.: Labor Praxis, *April*, 310 (1982)
44. Kaiser, R. E. et al.: Instrumental High Performance Thin-Layer Chromatography (Kaiser, R. E. Ed.), Bad Dürkheim 1982
45. Kaiser, R. E.: J. High Resol. Chrom., *Sept.*, 164 (1978)
46. Jänchen, D. E.: Instrumental HPTLC (Kaiser, R. E. et al. Ed.), Hüthig Verlag, Heidelberg, Basel, New York 1980
47. Rücker, G., Neugebauer, M., Mohie Sharaf El Din: Planta med. *43*, 299 (1981)
48. Amin, M., Jakobs, U.: J. Chromatogr. *131*, 391 (1977)
49. Grebian, B., Geissler, H. E., Mutschler, E.: Arzneimittelforschung *26*, 2125 (1976)
50. Lawrence, J. F., Frei, R. W.: J. Chromatogr. *79*, 223 (1973)
51. Ritter, W.: Instrumental High Performance Thin-layer Chromatography (Kaiser, R. E. Ed.), Bad Dürkheim 1982
52. Kreuzig, F.: GIT Fachz. Lab. Supplement Chromatographic, Ch 46 (1982)
53. Verpoorte, R., Baerheim Svendsen, A.: Zbl. Pharm. *118*, 563 (1979)
54. Jork, H., Wimmer, H.: Literatursammlung Quantitative Auswertung von Dünnschichtchromatogrammen, GIT Verlag Ernst Giebeler, Darmstadt 1982
55. Grimm, W.: Pharm. Ztg. *125*, 712 (1980)
56. Jänchen, D. E.: Instrumental High Performance Thin-layer Chromatography (Kaiser, R. E. Ed.), Bad Dürkheim 1982
57. Schwedt, G.: Chromatographische Methoden in der anorganischen Analytik, Hüthig Verlag Heidelberg, Basel, New York 1980
58. Grimm, W., Schepky, G.: Stabilitätsprüfung in der Pharmazie, Editio Cantor Aulendorf 1980
59. Tyihak, E., Szekely, T. J., Mincsovics, E.: Instrumental High Performance Thin-layer Chromatography (Kaiser, R. E. Ed.), Bad Dürkheim 1982

Author Index Volumes 101–126

Contents of Vols. 50–100 see Vol. 100
Author and Subject Index Vols. 26–50 see Vol. 50

The volume numbers are printed in italics

Anders, A.: Laser Spectroscopy of Biomolecules, *126*, 23–49 (1984).
Ashe, III, A. J.: The Group 5 Heterobenzenes Arsabenzene, Stibabenzene and Bismabenzene. *105*, 125–156 (1982).
Austel, V.: Features and Problems of Practical Drug Design, *114*, 7–19 (1983).

Balaban, A. T., Motoc, I., Bonchev, D., and Mekenyan, O.: Topilogical Indices for Structure-Activity Correlations, *114*, 21–55 (1983).
Baldwin, J. E., and Perlmutter, P.: Bridged, Capped and Fenced Porphyrins. *121*, 181–220 (1984).
Barkhash, V. A.: Contemporary Problems in Carbonium Ion Chemistry I. *116/117*, 1–265 (1984).
Barthel, J., Gores, H.-J., Schmeer, G., and Wachter, R.: Non-Aqueous Electrolyte Solutions in Chemistry and Modern Technology. *111*, 33–144 (1983).
Barron, L. D., and Vrbancich, J.: Natural Vibrational Raman Optical Activity. *123*, 151–182 (1984)
Bestmann, H. J., Vostrowsky, O.: Selected Topics of the Wittig Reaction in the Synthesis of Natural Products. *109*, 85–163 (1983).
Beyer, A., Karpfen, A., and Schuster, P.: Energy Surfaces of Hydrogen-Bonded Complexes in the Vapor Phase. *120*, 1–40 (1984).
Böhrer, I. M.: Evaluation Systems in Quantitative Thin-Layer Chromatography, *126*, 95–118 (1984).
Boekelheide, V.: Syntheses and Properties of the $[2_n]$ Cyclophanes, *113*, 87–143 (1983).
Bonchev, D., see Balaban, A. T., *114*, 21–55 (1983).
Bourdin, E., see Fauchais, P.: *107*, 59–183 (1983).

Charton, M., and Motoc, I.: Introduction, *114*, 1–6 (1983).
Charton, M.: The Upsilon Steric Parameter Definition and Determination, *114*, 57–91 (1983).
Charton, M.: Volume and Bulk Parameters, *114*, 107–118 (1983).
Chivers, T., and Oakley, R. T.: Sulfur-Nitrogen Anions and Related Compounds. *102*, 117–147 (1982).
Consiglio, G., and Pino, P.: Asymmetrie Hydroformylation. *105*, 77–124 (1982).
Coudert, J. F., see Fauchais, P.: *107*, 59–183 (1983).

Dyke, Th. R.: Microwave and Radiofrequency Spectra of Hydrogen Bonded Complexes in the Vapor Phase. *120*, 85–113 (1984).

Ebel, S.: Evaluation and Calibration in Quantitative Thin-Layer Chromatography, *126*, 71–94 (1984).
Edmondson, D. E., and Tollin, G.: Semiquinone Formation in Flavo- and Metalloflavoproteins. *108*, 109–138 (1983).
Eliel, E. L.: Prostereoisomerism (Prochirality). *105*, 1–76 (1982).

Fauchais, P., Bordin, E., Coudert, F., and MacPherson, R.: High Pressure Plasmas and Their Application to Ceramic Technology. *107*, 59–183 (1983).

Fujita, T., and Iwamura, H.: Applications of Various Steric Constants to Quantitative Analysis of Structure-Activity Relationshipf, *114*, 119–157 (1983).

Gerson, F.: Radical Ions of Phanes as Studied by ESR and ENDOR Spectroscopy. *115*, 57–105 (1983).
Gielen, M.: Chirality, Static and Dynamic Stereochemistry of Organotin Compounds. *104*, 57–105 (1982).
Gores, H.-J., see Barthel, J.: *111*, 33–144 (1983).
Green, R. B.: Laser-Enhanced Ionization Spectroscopy, *126*, 1–22 (1984).
Groeseneken, D. R., see Lontie, D. R.: *108*, 1–33 (1983).
Gurel, O., and Gurel, D.: Types of Oscillations in Chemical Reactions. *118*, 1–73 (1983).
Gurel, D., and Gurel, O.: Recent Developments in Chemical Oscillations. *118*, 75–117 (1983).
Gutsche, C. D.: The Calixarenes. *123*, 1–47 (1984).

Heilbronner, E., and Yang, Z.: The Electronic Structure of Cyclophanes as Suggested by their Photoelectron Spectra. *115*, 1–55 (1983).
Hellwinkel, D.: Penta- and Hexaorganyl Derivatives of the Main Group Elements. *109*, 1–63 (1983).
Hess, P.: Resonant Photoacoustic Spectroscopy. *111*, 1–32 (1983).
Hilgenfeld, R., and Saenger, W.: Structural Chemistry of Natural and Synthetic Ionophores and their Complexes with Cations. *101*, 3–82 (1982).
Holloway, J. H., see Selig, H.: *124*, 33–90 (1984).

Iwamura, H., see Fujita, T., *114*, 119–157 (1983).

Jørgensen, Ch. K.: The Problems for the Two-electron Bond in Inorganic Compounds, *124*, 1–31 (1984).

Kaden, Th. A.: Syntheses and Metal Complexes of Aza-Macrocycles with Pendant Arms having Additional Ligating Groups. *121*, 157–179 (1984).
Karpfen, A., see Beyer, A.: *120*, 1–40 (1984).
Káš,J., Rauch, P.: Labeled Proteins, Their Preparation and Application. *112*, 163–230 (1983).
Keat, R.: Phosphorus(III)-Nitrogen Ring Compounds. *102*, 89–116 (1982).
Kellogg, R. M.: Bioorganic Modelling — Stereoselective Reactions with Chiral Neutral Ligand Complexes as Model Systems for Enzyme Catalysis. *101*, 111–145 (1982).
Kniep, R., and Rabenau, A.: Subhalides of Tellurium. *111*, 145–192 (1983).
Krebs, S., Wilke, J.: Angle Strained Cycloalkynes. *109*, 189–233 (1983).
Kobayashi, Y., and Kumadaki, I.: Valence-Bond Isomer of Aromatic Compounds. *123*, 103–150 (1984).
Koptyug, V. A.: Contemporary Problems in Carbonium Ion Chemistry III Arenium Ions — Structure and Reactivity. *122*, 1–245 (1984).
Kosower, E. M.: Stable Pyridinyl Radicals. *112*, 117–162 (1983).
Kumadaki, I., see Kobayashi, Y.: *123*, 103–150 (1984).

Laarhoven, W. H., and Prinsen, W. J. C.: Carbohelicenes and Heterohelicenes, *125*, 63—129 (1984).
Labarre, J.-F.: Up to-date Improvements in Inorganic Ring Systems as Anticancer Agents. *102*, 1–87 (1982).
Laitinen, R., see Steudel, R.: *102*, 177–197 (1982).
Landini, S., see Montanari, F.: *101*, 111–145 (1982).
Lavrent'yev, V. I., see Voronkov, M. G.: *102*, 199–236 (1982).
Lontie, R. A., and Groeseneken, D. R.: Recent Developments with Copper Proteins. *108*, 1–33 (1983).
Lynch, R. E.: The Metabolism of Superoxide Anion and Its Progeny in Blood Cells. *108*, 35–70 (1983).

McPherson, R., see Fauchais, P.: *107*, 59–183 (1983).
Majestic, V. K., see Newkome, G. R.: *106*, 79–118 (1982).
Manabe, O., see Shinkai, S.: *121*, 67–104 (1984).
Margaretha, P.: Preparative Organic Photochemistry. *103*, 1–89 (1982).
Martens, J.: Asymmetric Syntheses with Amino Acids. *125*, 165—246 (1984).
Matzanke, B. F., see Raymond, K. N.: *123*, 49–102 (1984).
Mekenyan, O., see Balaban, A. T., *114*, 21–55 (1983).
Montanari, F., Landini, D., and Rolla, F.: Phase-Transfer Catalyzed Reactions. *101*, 149–200 (1982).
Motoc, I., see Charton, M.: *114*, 1–6 (1983).
Motoc, I., see Balaban, A. T.: *114*, 21–55 (1983).
Motoc, I.: Molecular Shape Descriptors. *114*, 93–105 (1983).
Müller, F.: The Flavin Redox-System and Its Biological Function. *108*, 71–107 (1983).
Müller, G., see Raymond, K. N.: *123*, 49–102 (1984).
Müller, W. H., see Vögtle, F.: *125*, 131—164 (1984).
Murakami, Y.: Functionalited Cyclophanes as Catalysts and Enzyme Models. *115*, 103–151 (1983).
Mutter, M., and Pillai, V. N. R.: New Perspectives in Polymer-Supported Peptide Synthesis. *106*, 119–175 (1982).

Naemura, K., see Nakazaki, M.: *125*, 1–25 (1984).
Nakazaki, M., Yamamoto, K., and Naemura, K.: Stereochemistry of Twisted Double Bond Systems, *125*, 1–25 (1984).
Newkome, G. R., and Majestic, V. K.: Pyridinophanes, Pyridinocrowns, and Pyridinycryptands. *106*, 79–118 (1982).

Oakley, R. T., see Chivers, T.: *102*, 117–147 (1982).

Painter, R., and Pressman, B. C.: Dynamics Aspects of Ionophore Mediated Membrane Transport. *101*, 84–110 (1982).
Paquette, L. A.: Recent Synthetic Developments in Polyquinane Chemistry. *119*, 1–158 (1984)
Perlmutter, P., see Baldwin, J. E.: *121*, 181–220 (1984).
Pillai, V. N. R., see Mutter, M.: *106*, 119–175 (1982).
Pino, P., see Consiglio, G.: *105*, 77–124 (1982).
Pommer, H., Thieme, P. C.: Industrial Applications of the Wittig Reaction. *109*, 165–188 (1983).
Pressman, B. C., see Painter, R.: *101*, 84–110 (1982).
Prinsen, W. J. C., see Laarhoven, W. H.: *125*, 63–129 (1984).

Rabenau, A., see Kniep, R.: *111*, 145–192 (1983).
Rauch, P., see Káš, J.: *112*, 163–230 (1983).
Raymond, K. N., Müller, G., and Matzanke, B. F.: Complexation of Iron by Siderophores A Review of Their Solution and Structural Chemistry and Biological Function. *123*, 49–102 (1984).
Recktenwald, O., see Veith, M.: *104*, 1–55 (1982).
Reetz, M. T.: Organotitanium Reagents in Organic Synthesis. A Simple Means to Adjust Reactivity and Selectivity of Carbanions. *106*, 1–53 (1982).
Rolla, R., see Montanari, F.: *101*, 111–145 (1982).
Rossa, L., Vögtle, F.: Synthesis of Medio- and Macrocyclic Compounds by High Dilution Principle Techniques, *113*, 1–86 (1983).
Rzaev, Z. M. O.: Coordination Effects in Formation and Cross-Linking Reactions of Organotin Macromolecules. *104*, 107–136 (1982).

Saenger, W., see Hilgenfeld, R.: *101*, 3–82 (1982).
Sandorfy, C.: Vibrational Spectra of Hydrogen Bonded Systems in the Gas Phase. *120*, 41–84 (1984).
Schlögl, K.: Planar Chiral Molecural Structures, *125*, 27–62 (1984).

Schmeer, G., see Barthel, J.: *111*, 33–144 (1983).
Schöllkopf, U.: Enantioselective Synthesis of Nonproteinogenic Amino Acids. *109*, 65–84 (1983).
Schuster, P., see Beyer, A., see *120*, 1–40 (1984).
Schwochau, K.: Extraction of Metals from Sea Water, *124*, 91–133 (1984).
Selig, H., and Holloway, J. H.: Cationic and Anionic Complexes of the Noble Gases, *124*, 33–90 (1984).
Shibata, M.: Modern Syntheses of Cobalt(III) Complexes. *110*, 1–120 (1983).
Shinkai, S., and Manabe, O.: Photocontrol of Ion Extraction and Ion Transport by Photofunctional Crown Ethers. *121*, 67–104 (1984).
Shubin, V. G.: Contemporary Problems in Carbonium Ion Chemistry II. *116/117*, 267–341 (1984).
Siegel, H.: Lithium Halocarbenoids Carbanions of High Synthetic Versatility. *106*, 55–78 (1982).
Sinta, R., see Smid, J.: *121*, 105–156 (1984).
Smid, J., and Sinta, R.: Macroheterocyclic Ligands on Polymers. *121*, 105–156 (1984).
Steudel, R.: Homocyclic Sulfur Molecules. *102*, 149–176 (1982).
Steudel, R., and Laitinen, R.: Cyclic Selenium Sulfides. *102*, 177–197 (1982).
Suzuki, A.: Some Aspects of Organic Synthesis Using Organoboranes. *112*, 67–115 (1983).
Szele, J., Zollinger, H.: Azo Coupling Reactions Structures and Mechanisms. *112*. 1–66 (1983).

Tabushi, I., Yamamura, K.: Water Soluble Cyclophanes as Hosts and Catalysts, *113*, 145–182 (1983).
Takagi, M., and Ueno, K.: Crown Compounds as Alkali and Alkaline Earth Metal Ion Selective Chromogenic Reagents. *121*, 39–65 (1984).
Takeda, Y.: The Solvent Extraction of Metal Ions by Crown Compounds. *121*, 1–38 (1984).
Thieme, P. C., see Pommer, H.: *109*, 165–188 (1983).
Tollin, G., see Edmondson, D. E.: *108*, 109–138 (1983).

Ueno, K. see Takagi, M.: *121*, 39–65 (1984).

Veith, M., and Recktenwald, O.: Structure and Reactivity of Monomeric, Molecular Tin(II) Compounds. *104*, 1–55 (1982).
Venugopalan, M., and Vepřek, S.: Kinetics and Catalysis in Plasma Chemistry. *107*, 1–58 (1982).
Vepřek, S., see Venugopalan, M.: *107*, 1–58 (1983).
Vögtle, F., see Rossa, L.: *113*, 1–86 (1983).
Vögtle, F.: Concluding Remarks. *115*, 153–155 (1983).
Vögtle, F., Müller, W. M., and Watson, W. H.: Stereochemistry of the Complexes of Neutral Guests with Neutral Crown Host Molecules, *125*, 131–164 (1984).
Volkmann, D. G.: IonPair Chromatography on Reversed-Phase Layers *126*, 51–69 (1984).
Vostrowsky, O., see Bestmann, H. J.: *109*, 85–163 (1983).
Voronkov. M. G., and Lavrent'yev, V. I.: Polyhedral Oligosilsequioxanes and Their Homo Derivatives. *102*, 199–236 (1982).
Vrbancich, J., see Barron, L. D.: *123*, 151–182 (1984).

Wachter, R., see Barthel, J.: *111*, 33–144 (1983).
Watson, W. H., see Vögtle, F.: *125*, 131–164 (1984).
Wilke, J., see Krebs, S.: *109*, 189–233 (1983).

Yamamoto, K., see Nakazaki, M.: *125*, 1–25 (1984).
Yamamura, K., see Tabushi, I.: *113*, 145–182 (1983).
Yang, Z., see Heilbronner, E.: *115*, 1–55 (1983).

Zollinger, H., see Szele, I.: *112*, 1–66 (1983).

Bei Fragen zur Produktsicherheit wenden Sie sich bitte an:
If you have any questions regarding product safety,
please contact:

Walter de Gruyter GmbH
Genthiner Straße 13
10785 Berlin
productsafety@degruyterbrill.com